The Homeowner's Guide to Carpentry and Cabinetry

The Homeowner's Guide to Carpentry and Cabinetry

By
K. E. Armpriester
and
B. A. Bremer

Grolier Book Clubs, Inc.
Danbury, Connecticut

Published by
   Popular Science Books
   Grolier Book Clubs
   Sherman Turnpike
   Danbury, CT 06816

Book design by Linda Watts

Produced by Bookworks, Inc., West Milton, Ohio
Layout Artist: Linda Ball
Illustrators: Mary Jane Favorite
              Christine Vogel
Cover Photography: © Mark D. Thellmann 1989
How-to Photography: Kate Armpriester
                    Beverly Bremer
                    Mark Bremer
                    Karen Callahan
Typesetting: Computer Typography, Huber Heights, Ohio
Proofreader: Hue Park
Editorial Consultant: Nick Engler
Editorial Assistant: Chris Walendzak

Many thanks to:

Artistic Homes, Inc., Englewood, Ohio
Fultz Builders, Englewood, Ohio
H & M Construction Co., Inc., Union, Ohio
LanCo Construction and Development, Inc.,
   St. Petersburg, Florida
Scott Laubach, Largo, Florida
Don Myers Electric, Inc., Vandalia, Ohio
R.F. Pitsenbarger Construction Co., Inc., Troy, Ohio
Rarick Homes, Inc., Troy, Ohio
Supreme Industries, Dayton, Ohio

Library of Congress Cataloging-in-Publication Data

Armpriester, K.E.
   The homeowner's guide to carpentry and cabinetry/
   by K.E. Armpriester and B.A. Bremer
   p.   cm.
   ISBN: 1-55654-059-0
   1. Carpentry   2. Cabinet-work
   I. Bremer, B.A.   II. Title
     III. Title: The homeowner's guide to carpentry and
       cabinetry
             1989

Contents

Introduction.. vi
Tools, Materials, and Techniques
    1. Getting Started 1
    2. The Right Tools 9
    3. Woodworking Materials and Hardware............................25
    4. Basic Woodworking Skills 39
Home Carpentry
    5. Basic Residential Construction 59
    6. The Fundamentals of Framing..77
    7. Roofing and Siding 103
    8. Finishing Touches 123
Home Cabinetry
    9. Basic Cabinet Construction..143
    10. Installing Cabinets 173
Special Additions
    11. Doors and Windows...185
    12. Staircases and Railings 219
    13. Open Decks...229
Appendices
    Estimating Construction Materials 240
    Estimating Cabinetmaking Materials...................................240
    Budgeting Carpentry and Cabinetry Projects 241
Glossary of Carpentry and Cabinetry Terms 242
Index 245

Introduction:
In Pursuit of Dream Homes

We think of home as a place of significance, substance, and permanence: It will always be there for us, as it has been all along. In reality, however, home is a place of growth and change. The building may stay in one place, but the needs of its occupants constantly shift. Consequently, any given room or area may be transformed over and over as family members mature and progress.

For this reason, your "dream home" is just that — an elusive idea that you can never quite make a reality. As fast as you build or remodel the house to satisfy the needs of the family, those needs change — and you must build and remodel some more. For the homeowner, planning and building becomes an eternal cycle, as sure as day and night.

This book will guide you and help you in the never-ending pursuit of your dream home. It won't help you achieve a dream home — authors Armpriester and Bremer warn you to have no illusions on that score. However, it will make the chase more fruitful and pleasant.

Beginning at the Beginning

Before you jump into this book — and another planning/building cycle — take a few moments to reflect on the first part of the process, the *planning* part. Many do-it-yourselfers give short shrift to this step. With book in hand, they rush into a building or remodeling project. Halfway through, they discover the results might have been better if they'd just thought out the project more carefully. To help avoid this, here are a few considerations to mull over *before* you cut your first board:

Define your goals. The first step in a successful home carpentry or cabinetry project is to know exactly why you're tackling the project in the first place. Sounds funny, doesn't it? You'd be surprised how many people never give this adequate deliberation. As a result, they often spend more in time, money, and materials than needed to accomplish their goals.

Suppose, for instance, that you decide to build new kitchen cabinets. Ask yourself *why* you need them. What are your true goals? Do you wish to create more storage? Do you want to make the arrangement of appliances and cabinets more efficient? Do you want to dress up an old kitchen? If you need more storage, a new set of cabinets may only be a partial solution; you may also need to expand the kitchen. If the present arrangement is inefficient, you should redesign the kitchen completely. This may involve new plumbing and electrical runs in addition to new cabinets. If you're tired of the way your old kitchen looks, you may not need to replace the cabinets at all. Just build new doors and drawer fronts. Once you define the goals, it becomes clearer what you must do to accomplish them.

Write down these goals. It's much easier to keep them in focus when you can see them in front of you. List them in order of importance — you may not have the money or the time to accomplish everything right now. Finally, check them with other members of your household. They may want to make additions or amendments to your list.

Have a long range plan. Most homeowners aren't just one or two projects away from their dream home. We usually have a dozen or more that we want to tackle, eventually. Write these down and consider the order in which they ought to be done. *Try not to think about the order of importance!* Just plan a logical sequence.

For example, you may want to add another bedroom and a deck onto the back of your home — two separate projects. The bedroom is a major project; it will require a lot of effort and money. The deck is minor and more affordable. The natural tendency is to build the deck first — you can enjoy the extra space while you're saving up for the bedroom. Once built, however, the deck might be in the way of the bedroom project. You may even have to remove part of the deck to build the bedroom. To avoid wasting time and money, do things in their proper order.

Your long range plan doesn't have to be cast in concrete, of course. It can (and most probably will) change as your family grows. But although you revise it from time to time, a general plan gives you a measuring stick to see how close you're getting to your dream home. If you do projects helter-skelter, you never know how close you are.

Don't over-invest. There isn't a homeowner alive who couldn't think of a way to spend a million dollars and half a lifetime building a dream home. But is it worth it?

When deciding whether or not to invest the money and the time in building and remodeling projects, consider this: How long do you intend to live in your present home? You may have found the house and the neighborhood in which you want to spend the rest of your life. If so, it won't matter how much money or time a project requires, so long as it makes your home more pleasant.

However, many homeowners move every few years. You may want to sell your home eventually and realize a return on your investment. If so, consider how much each project will add to the value of your home. Most projects — new kitchen cabinets, decks, garages and carports — will give you an even return. A few, such as an extra bathroom, will actually pay you more. However, some won't pay as much as they cost. A greenhouse, for instance, will not add much to the value of your real estate.

Also consider the average cost of houses in your neighborhood. Location has a good deal to do with the value of real estate, and houses usually don't bring too much more than the local norm, no matter how extensively they've been remodeled. If you plan to sell your home some day, you don't want to put more money into it than you can recover. If you have any doubts whether or not you can recoup the cost of a project when it's time to sell and move on, check with a local real estate agent.

Know your limitations. No matter how experienced you may be, there are limits to what you can do. These limits are determined not only by your skills and expertise, but also by the time, tools, and materials available to you. When you plan a project, be honest with yourself. Don't take on a task that's beyond your ability or resources to complete it.

There's no quicker way to get sick and tired of doing it yourself than to bite off more than you can comfortably chew.

If you're a novice, start with small projects and work your way into bigger ones. This book will help you expand your skills and tackle carpentry and cabinetry jobs you've never done before, but don't try to do too much too quickly. It's difficult to build a complete set of kitchen cabinets if you've never hung a door or installed a drawer. Small successes build confidence, patience, and the expertise needed to tackle larger projects.

And remember: Although a good building or remodeling job will add to the value of your real estate, a botched job may detract. If you're unsure about how to go about a particular job, don't hesitate to hire some experienced help. You can learn from watching as well as doing. Also remember that no matter how experienced you become, there will always be some jobs that are best left to professionals. These will vary from individual to individual, depending on experience and aptitude. In general, however, major plumbing, electrical, and concrete work should be done by people who are properly trained for it.

Draw up a set of plans. Before you tackle any carpentry or cabinetry project, large or small, make a set of plans. You don't have to have any artistic ability to do this; these don't have to be architectural-quality blueprints. Usually, all that's needed is a sketch with overall measurements. Whether you make them simple or fancy, plans are important for several reasons:

First, they help you to visualize the completed project. By making a sketch, you can see the proportions of a new addition and how it will look in place. This way, you can make changes before you begin.

Second, plans help you to estimate materials. As you put the dimensions down on paper, you can begin to list the sizes and quantities of the building materials you will need. This, in turn, helps to eliminate time-consuming trips to the hardware or lumber store.

Finally, and most importantly, plans help you to think through a project before you actually do it. To draw a project on paper, you must mentally build it. As you do this, you devise ways to accomplish certain steps. Modifications will occur to you that make the project more useful, less expensive, and easier to build. In the end, you'll get much better results — and a little closer to your dream home.

Tools,
Materials,
and
Techniques

1

Getting Started

When you pass a construction site are you amazed at the speed and precision at which structures take shape? Do you envy the architect whose ideas start it all, the experienced carpenter who with his hammer and saw forms the framework, or the cabinetmaker whose talents add beauty and grace? Maybe you've thought about tackling projects that call for a combination of these skills — on a smaller scale of course. At least at first.

Why not start with an addition to your home? You can make it as simple or as intricate as your skills, confidence and time will allow. Building any type of structure, whether it is a garage, family room, storage shed, or kitchen, is done one step at a time. Therefore — you can do it. Let your imagination run wild — be the architect, the carpenter and the cabinetmaker. You can add your own special touches, and build what *you* want. You no longer have to buy off the contractor's "rack", that is to say you're the boss.

Whether you do all or part of the work yourself, you'll learn new skills, gain confidence, save money, add value to your home and create extra living space for your family. But the best reward of all — when you pass those work sites, you'll be able to say to yourself, "I've done that. It was hard work, but it was also a lot of fun."

Let's Get Busy

Now that you've decided to build an addition to your home, what do you do first?

This chapter and the next three will highlight how to get started, what tools you'll need, what materials you'll use, and how to master basic woodworking skills. The following chapters will detail the actual steps you'll take.

First, decide what type of addition you want, its size, and how much you want to spend. Check local building codes and find out what permits are needed. Then sketch your plans. You'll probably want to have an architect check to see if they're acceptable for submission to sub-contractors and for application for permits.

Next, determine who will do the work. You have several choices. Do all the work yourself. Or better yet, you can act as the contractor — do part of the work yourself and hire sub-contractors to do the rest. The big advantage to using sub-contractors is that they are familiar with local codes. Foundation work,

SAFETY CHECKLIST

1 Before you plug in your power tool, be sure to check the output and the amperage of the circuit you'll be using. The output of the circuit **must** meet or exceed the electrical requirements of the tool. The amperage must be rated high enough to handle the load (in amps) of that motor, plus any other tools or equipment you may have plugged into the same circuit. Typically, the minimum rating should be 15 amps, but a 20-amp circuit will provide you with an even greater safety margin.

2 Power tools can be hazardous; be sure to read the manufacturer's instructions and precautions before using them. Keep the owner's manuals and instructions that came with your tools in a notebook or other convenient location.

3 Use only power tools that are **properly grounded.** These will be identified as having three-prong plugs. **NEVER MODIFY A PLUG; if your plug will not fit an outlet, have the proper outlet installed.**

4 If you must use extension cords, be sure they have 3-prong plugs. When choosing an extension cord for a power tool, make sure that you have selected the proper wire gauge for the tool. The longer the cord, the thicker it must be. For example, a power tool that requires 14 amps should have a No. 14 wire gauge if it measures 25 feet in length. If the cord is 100 feet long, then the wire gauge will need to be increased to No. 10. Tool manufacturers usually supply this information with their products.

5 Equip your work area with adequate lighting. Poor lighting increases the risk of accidents. Fluorescent lighting is a good choice since it won't produce glare. (Too *much* light can also be dangerous.) If you're doing a lot of fine and close woodworking, invest in a portable drafting lamp which can be directed to your work. If you're doing carpentry work outdoors, be careful of glare from the sun and be sure to stop well in advance of nightfall.

6 Dress for the job. Remove all jewelry and your watch. Wear short-sleeved work shirts or roll up your sleeves above the elbows. Do not wear gloves; they can easily catch on the moving parts of power tools. (Exceptions: If you're doing only the fitting and constructing tasks of carpentry, you may wear work gloves for protection against splinters. You may also wear rubber gloves when applying adhesives or solvents.) Wear proper non-slip footwear — leather and rubber-soled shoes for cabinetmaking, steel-toed shoes for carpentry. If your hair is long, secure it with pins or wear a hat to keep it out of your way. In some instances, it may be necessary to wear a hard hat.

7 Protect your eyes. Wear safety glasses, goggles, or a face shield.

8 As a safeguard against hearing loss when operating a power tool such as a router that produces high pitch tones, always wear hearing protectors.

plumbing, wiring and installing drywall are tasks you may choose to have someone else accomplish. Also you might have a cabinetmaker make the cabinets, then you could design and make the doors. Since subcontractors are usually willing to share some of the tricks of their trade, ask questions and pay attention to what they do. You might learn enough so you can do it yourself the next time.

Before you decide who will do the work, make detailed lists of all the tasks. Also evaluate your skills and be realistic about the time you will have to devote to the project.

9 Wear a dust mask during sanding operations — or any operation that creates fine dust. When working with any harmful substances such as finishes or solvents, wear a respirator or painter's mask. Work in a well ventilated space. Also, to avoid problems, take frequent breaks. Smoking is strictly forbidden in the workshop setting because of the danger of fire or explosion. If you must smoke, leave your work area.

10 Keep work areas clean and free of clutter to avoid accidents. Clean up as often as possible; be sure to pick up bent nails and wood scraps, and clean up spills. Avoid working with helpers in cramped quarters where another person's tool or materials could harm you.

11 If you're breaking into walls for built-ins or to hang cabinets, be cautious. Don't hammer, drill, saw or work in any areas where you suspect electrical wiring or plumbing might exist. Don't guess — be sure!

12 Stay organized and calm. Have a place for everything and everything in its place. Not only will your work area be safer, but your work will go smoother and faster and be more accurate. Do not work if you're nervous, tired, angry or under the influence of alcohol or drugs; these states can cause carelessness. The chances of an accident are slimmer if you're relaxed, rested, alert and able to fully concentrate on your work.

13 Keep children and visitors a safe distance from the work area.

Safe and Sound Practices

Safety is absolutely critical to any project where power tools are used. Power tools are extremely dangerous when not used properly; therefore you and (if you have one) your helper, must always follow all safety procedures.

Woodworking safety involves not only how you use each tool, but also the condition and organization of your workplace, the lighting, your clothing, even your state of mind. Throughout this section, special **CAUTIONS** and safety tips will be given for specific tasks. In addition, read and understand the 'Safety Checklist':

Special Gear for Eyes, Ears, and Lungs. Obviously, eyes need special protection during power and hand tool operations where there is danger to your eyes. When choosing equipment, you should consider your own safety first, and physical comfort next. Safety glasses will give you 'general' protection from large objects because they have shatter-resistant lenses. However, safety glasses do not *completely* cover the eyes, as do goggles. Goggles are an excellent choice and some are designed to be worn over regular prescription glasses. If you prefer even greater protection, consider using a face shield. This apparatus is easily raised and lowered as you need it.

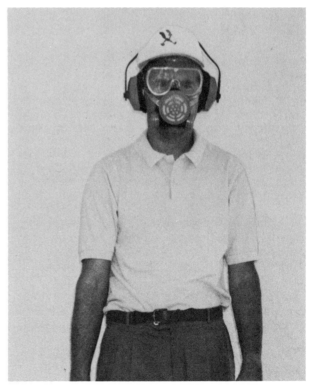

This woodworker is properly dressed for the workshop. He is wearing proper eye and ear protection, no jewelry or gloves, a short-sleeved shirt, and leather shoes with rubber soles. In addition, he will tuck long hair under a cap and when performing operations that generate a lot of fine or toxic sawdust he will wear a dust mask or respirator. If he were doing structural tasks, the same outfit would be worn except the shoes would be steel-toed.

Ears also deserve special treatment; yet this safety practice is often neglected. The sound of power tools in operation, and even nails being pounded in confined areas, can be damaging to your ears. You probably won't notice it, since hearing loss is subtle — it takes place very slowly. Earmuff-type protectors are excellent for filtering out the noise that you don't need but allowing you to still hear your helpers or someone else calling out to you.

Don't ignore your lungs. Sawdust particles are harmful enough, but when you consider all the chemicals that woodworking can and often includes, protecting your lungs becomes a major issue. To shield your breathing passages and lungs, you should wear a quality dust mask or respirator. As you use the apparatus, be sure to change the filters as frequently as the manufacturer recommends.

Power Tool Danger Zones. Because power tools are so potentially hazardous, there are certain places on and around them that should be treated with respect; these are the areas where accidents most often occur. These zones are appropriately called *danger zones*.

Each power tool will have its own danger zone, and this information should be clearly supplied by the manufacturer. The danger zones shown on page 5 are for two of the most common (and most common sizes of) woodworking tools — the table saw and the radial arm saw. Study the diagrams and then be sure to observe the following when working close to the danger zones:

■ Always keep fingers, hands, and all parts of your body out of the danger zone when the machine is in operation.
■ Never reach into the danger zone to clear away chips or sawdust while the machine is running. Turn off the machine and let it come to a complete stop first.
■ Before reaching into the danger zone to make adjustments or mount accessories, turn off and unplug the machine. Before you plug the machine back in, make sure that the switch is still set to off.

Solutions for Sawdust. Though sawdust is a sign of progress for the woodworker, it can become a nuisance and a threat if it is not dealt with properly. Piles and trails of sawdust can become a fire hazard, plus it could cause you to slip and fall, no matter what kind of shoes you're wearing. Another problem is the risk to lungs and breathing passages; this is especially true for people with allergies, emphysema, or asthmatic conditions.

You can wear a dust mask or respirator to protect yourself from breathing excessive amounts of sawdust. Keep in mind, however, that the fineness of the dust and the accumulated volume in your workshop are crucial factors. It also helps to open a window or use an exhaust fan to ventilate your shop.

Besides using a broom faithfully, probably the best thing that you can do is to invest in a large shop vacuum or dust collection system. Such an apparatus is well worth the price, since it easily picks up dust

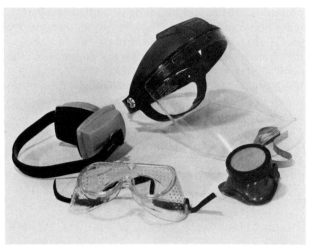

Shown clockwise from top: face shield, dust mask, goggles, and hearing protectors. *Face shields* cover the entire face and can be flipped up when not needed. *Dust masks* keep out sawdust and other fine particles while allowing you to breathe. *Goggles* offer more protection than safety glasses since they completely surround your eyes. By wearing *hearing protectors,* you'll guard against cumulative hearing loss. This specialized equipment screens out the damaging noise but still permits you to hear normal conversations.

from hard-to-get-to areas and saves you a lot of stooping and bending. You might even want to build your own collecting bin — such as a drawer beneath your table saw. If you're really adamant about keeping dust to a minimum, select power tools that have sawdust 'pickups' — receptacles for the attachment of a shop vacuum or dust collection system.

Setting Up a Work Space

The ideal situation for the woodworker, even at the beginner level, is to have a workshop — a closed area specifically created and maintained for the purpose of woodworking. With such a space, you can lock your door and keep all dangerous tools and materials from unauthorized visitors. If you don't have the luxury of a workshop, then you will need to be especially careful about where you locate your work space, who you permit to enter it, and how you store your supplies. If you only intend to do rough carpentry, you might not feel the need to set up a permanent workshop, since you'll mainly be working 'on site', setting up and breaking down your 'portable shop' as you go.

The process of planning for a workshop should focus on the tasks that you hope to accomplish in it, the tools you have (or hope to acquire), and the projects you want to do. Though we all have individual tastes and preferences, there are a few principles to keep in mind when working on your plan. First, decide how much space you can devote to a shop. Even if your space is small, with a little ingenuity, you can probably make it work for you. In any case, the tools will be close at hand.

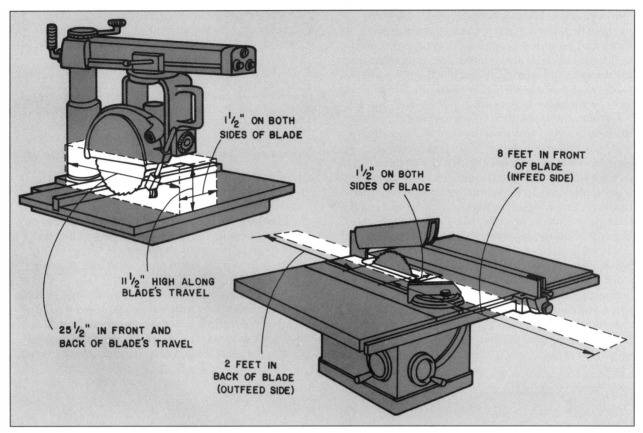

1 1/2" ON BOTH SIDES OF BLADE

1 1/2" ON BOTH SIDES OF BLADE

8 FEET IN FRONT OF BLADE (INFEED SIDE)

11 1/2" HIGH ALONG BLADE'S TRAVEL

25 1/2" IN FRONT AND BACK OF BLADE'S TRAVEL

2 FEET IN BACK OF BLADE (OUTFEED SIDE)

Every power tool is potentially dangerous. In addition to other safety precautions, **ALWAYS KEEP FINGERS, HANDS, AND ALL PARTS OF YOUR BODY OUT OF**

THE DANGER ZONE WHEN THE TOOL IS IN OPERATION. Shown are the danger zones for two commonly used power tools: the radial arm saw and table saw.

GETTING RID OF SAWDUST

This shop-made tray, positioned beneath the table saw, collects sawdust as it's being created. The dust stays in one place and is easily disposed of during cleanup.

The best way to control and collect sawdust in your shop is to connect a shop vacuum or a dust collection system to the dust chutes on your power tools.

Design for Success — The Work Triangle.
Most important in workshop design is the *work triangle*, a special triangular arrangement of your three key work areas. For most woodworkers, these are likely to be the workbench, the stationary power tools, and the storage area, where many hand tools and most materials are situated. The primary reason for using a work triangle is that it will save you steps. Generally, also, the triangular configuration will help to keep you more organized.

Once you have figured out what your three vital work stations are and what they will include, draw a scale plan on graph paper. Make paper cutouts, again to scale, of your workbench, power tools, storage shelves, racks, and any other large pieces you wish to place in the workshop. Move these around until you find a logical triangle. Keep the sides of the triangle as short as possible, but allow enough space between tools so that you won't feel cramped.

Arrange your tools around structures that can't be moved — doors, windows, water heaters, and ductwork. Or, if your garage is also your shop, design an arrangement that can be quickly set up and put away when you pull the car in and out. For stationary tools that require a lot of clearance, such as a table saw, make use of open doors and windows for the extra space you need.

Meeting Needs in the Workshop.   Once you begin to use your workshop, you'll recognize the need to make it a comfortable space. Some of the factors involved are the temperature, the 'furnishings', and the accessibility of tools.

If your shop is separate from your home, or if it's in an unheated area, you will probably need to supplement the heating system. In northern latitudes, keep your shop warm enough for comfort but cool enough for clear thinking. If you use a space heater,

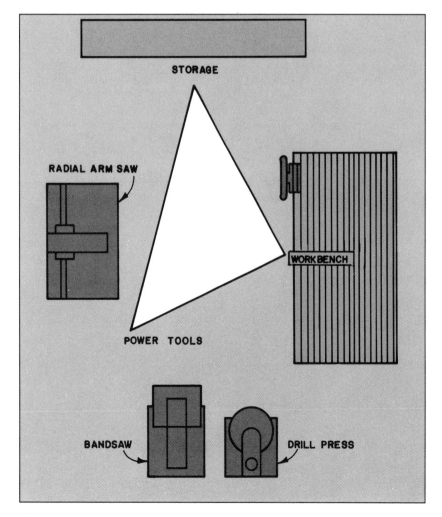

STORAGE

RADIAL ARM SAW

WORKBENCH

POWER TOOLS

BANDSAW   DRILL PRESS

The Work Triangle. Use this basic triangular pattern to arrange your workshop. By putting the workbench, power tools, and hand tool storage in three separate areas, you'll be well organized and save yourself steps. Make sure that the center area is small enough to be practical, but large enough to be safe and feel comfortable.

avoid exposed heating elements and open flames. For safety, you might consider using an oil-filled portable baseboard heater.

If you intend to use your workshop as the place where you work on project plans, consider furnishing it with a desk or drafting table. If you can't afford the furniture or the space, you can use your workbench for the same purpose. Simply find a comfortable stool or chair, and plan your workbench so you can slide your legs underneath when sitting.

Depending on what type of flooring you have in your workshop, you might want to 'pad the floor'. Concrete floors can be brutal to even the strongest feet and legs, so consider using 'anti-fatigue' mats in areas where you will stand often. Smaller mats can aid you in clean-up as you can simply pick them up and shake off the sawdust.

You should be very particular about where and how you store your tools — especially the ones that you will use frequently. The best rule is to keep them in plain sight and within easy reach. This is especially true of 'safety tools' such as push sticks and push blocks; these should be very accessible — right by the machine with which they're used.

Safety in the Workshop. No one likes to think of accidents happening, but the truth is that workshops are very dangerous areas and you should always be prepared. Use the following as a guide to prevent or quickly divert a workshop catastrophe:

■ When designing or remodeling a shop, use many separate electrical outlets; this will limit your use of trailing (and possibly tripping over) extension

■ When designing or remodeling a shop, use many separate electrical outlets; this will limit your use of trailing (and possibly tripping over) extension cords. If you must use extension cords, tape them to the floor. Never stand on them or run them under mats.

■ Store all flammable solvents in metal cans.

■ Make a fire extinguisher with the correct rating one of your essential tools; keep it in a central and easy-to-reach location.

■ Assemble or buy a first aid kit and keep it in a handy spot.

A Special Work Site. If you're working in an area far from your workshop, you'll need to choose a temporary work site where you will set up and break down tools on a regular, usually daily, basis. This work site could be indoors or outdoors.

Obviously, a factor in choosing a work site will be its proximity to the project, but barring this, there are other important considerations. Select an area that is relatively flat and affords you and your helpers plenty of space for movement. Also, find a place that's relatively free from traffic and passersby — both of which can be distractions.

The basic setup for a work site consists of a portable workbench, sawhorses, or a combination of both. A circular saw and numerous hand tools as mentioned in Chapter 2 should be within easy reach. If available, a table saw or radial arm saw would be helpful. Safety is, as always, essential and you should use the aforementioned 'safety checklist' as a general guide when setting up a work site. Because extension cords are usually the norm for work site situations,

The work site is often outdoors, as shown here. Although a table with a 'contractor's saw' is preferred, cutting may be done at sawhorses. Requirements for a good setup are adequate lighting, space, and electricity. When using power tools outdoors, be sure they are properly grounded and use extension cords that are recommended for outdoor use.

2

The Right Tools

In this chapter, you'll become familiar with some of the basic tools you'll use for your add-ons and built-ins. Here is everything from simple, traditional, and inexpensive hand tools to elaborate and costly stationary power tools.

When purchasing tools, we recommend a 'middle-of-the-road' approach for the beginning woodworker. Using all hand tools can be quite laborious, whereas investing in a whole workshop full of stationary tools is an expense you should put off until you're completely familiar with *your* needs. Instead, choose a mixture of basic hand tools and time-saving hand-held power tools such as a drill, sabre saw, circular saw, orbital sander, and a router. You'll be encouraged by the results using just these few tools.

If you're not a beginner, then use this chapter as a guide to purchase more advanced tools for use in your projects. Consider these factors: practicality (do you *really* need it?), price (can you afford it?), storage (do you have a workshop or place to store it?), and 'overlap' (can the same function be performed equally as well by another tool?).

When the time comes to purchase your new tool(s), shop very carefully. Ask friends who are woodworkers if they're satisfied with their brands. Is service readily available? Are replacement parts such as blades, belts and bits easy to get and are they compatible with other brands? In general, it's best to purchase medium- to high-priced tools. Cheap products can cause errors and frustrations. However, always compare features and benefits when shopping. Ask yourself what do you expect now and in the future from this tool. Once you've got the tools, you're ready to proceed with the training, which can range from simple practice with a hammer to taking a shop course on how to use a table saw.

Measuring and Marking Tools

Of utmost importance with any project is the process of measuring and marking. Structures that are not true are a disgrace. A mismarked and incorrectly cut member of a cabinet can cause extra expense, wasted time and much aggravation. For these reasons, always remember to: "Measure twice, cut once". And minimize your chances of error by using only quality tools and using them only for the job for which they were designed.

Tapes and Rules. A steel tape is bound to get a workout in any major project. Most are manufactured with locking buttons to aid you when working solo, plus they often have standard case measurements (such as 2 or 3 inches), to make it easier for you to calculate inside measurements. You might want to consider a type that has markings every 16 inches, the standard spacing for wall studs.

An old favorite is a folding wooden rule that offers rigidity which means that you can measure without support at the other end. These hinged devices are available in folded-up sizes of 6 or 8 inches. A helpful feature on some models is the sliding extension on the end which can be pulled out for inside measurements.

Squares, Bevels, Levels and Bobs. These tools are crucial for accuracy. Squares are essential for laying out and marking cutlines. All squares indicate 90° angles, and some also indicate angles of 30°, 45° and 60°. Be sure to purchase high-quality squares and take care not to drop them and bend them out of square. Periodically test your squares for trueness. Do this by holding the body against a straight-edged board and drawing a line along the blade. Flop the square and draw another line. The two lines should match precisely.

The sliding T-bevel, an adjustable tool, measures any angle between 0° and 180°. It may be used in conjunction with a protractor, or as a tool to match an existing angle.

Use a carpenter's level to gauge *level* and *plumb* (respectively, true horizontal and true vertical). This tool is indispensable when hanging cabinets and checking the components of structures.

The plumb bob uses gravity to accurately pinpoint true vertical. The plumb bob is used to transfer an overhead point to the surface below — or just the opposite. For example, if you're adding a partition wall to your present structure, align the plumb bob from above close to the point on the floor where the construction is to take place (but not *touching* the floor, as it should always swing freely before centering on the correct point).

Marking Instruments. Don't overlook the importance of accurate and clear markings. For straight lines, your best choice is a carpenter's pencil or a standard pencil with hard lead — one that won't require frequent sharpening. When the mark is especially critical, use a scratch awl. Remember, however, that awl marks can't be erased like pencil marks.

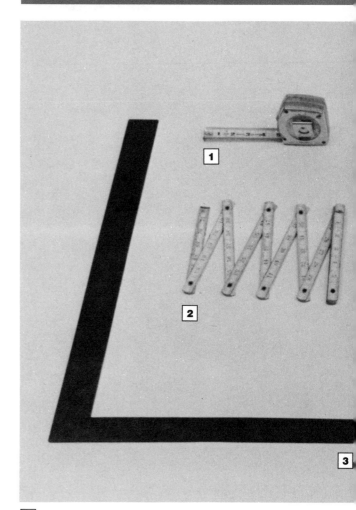

1 **Steel Tape.** Flexible steel tape with a locking button and end hook. The one inch or wider tape is sturdier and easier to use. Choose the length that best suits your needs. Some tapes have special markings every 16", the standard spacing for wall studs and floor joists.

2 **Folding Wooden Rule.** Made from rigid wooden sections hinged together; typically measures 6 feet in length. A brass sliding extension allows you to make accurate inside measurements.

3 **Carpenter's Square.** This steel straightedge has a 24-inch long side called a *body* and a 16-inch short side called a *tongue*. The 2-inch thick body meets the 1½-inch thick tongue at the *heel* of the square. Available with different kinds of information printed on them, for framing or installing rafters.

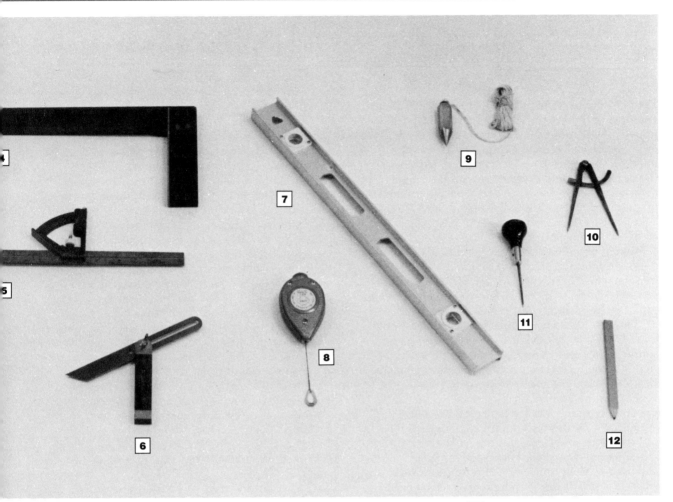

4 **Try Square.** Available with 6-, 8-, or 12-inch blades. For laying out cutoff lines across framing lumber or boards; also for checking squareness. Available with a bevel on the handle for short 45° angles.

5 **Combination Square.** Consists of a 12-inch blade and a sliding head with a 45° surface for miters and a spirit level for leveling operations. Often has a scribe in bottom of head for marking fine lines. Purchase a well-made, reputable brand.

6 **T-Bevel.** The handle pivots on the blade so any angle can be measured and transferred. Commonly available with an 8-inch blade.

7 **Carpenter's Level.** The standard level is made of wood or aluminum and is 24" long. It has three replaceable spirit level vials — center for level and ends for plumb. Some levels come with one end vial set at 45°.

8 **Chalk Line.** An easy means to mark large vertical or horizontal surfaces. A string coated with chalk is stretched taut between two points. By snapping the string, a mark is left on the surface. The chalk line is housed in a device that can double as a plumb bob.

9 **Plumb Bob.** A simple weighted fixture hanging by a string, this measuring tool is used to gauge and mark plumb.

10 **Dividers.** A hinged instrument with wings for stepping off measurements and scribing circles.

11 **Awl.** A sharp instrument for making precision marks; makes permanent impressions in wood.

12 **Pencil.** Select a pencil with hard lead, made specifically for woodworkers. Pencil marks can be erased, unlike awl marks.

Use a chalk line when you have a large area to mark. A chalk line is basically a length of string stretched taut between two points. Chalk is rubbed along the entire length of the string, then the string is snapped which transfers the chalk to the surface.

Another aid that is particularly good for marking curves and for 'repeat' jobs like stepping off small measurements is the hinged dividers.

Hand Tools — for Muscle and Mind

Though cutting, sanding, and many other functions are more and more being accomplished with the help of power tools, there are still many tasks that even the most elaborately equipped woodworker must and will want to perform by hand. Moreover, the beginner should pay attention to these hand tools because it is with them that the basic principles of woodworking are learned. These tools might take a little extra effort and time but they also give you a great learning experience.

Often, precision cutting or smoothing must be done carefully and accurately, so hand tools and frequent checking are the answer. Hand tools offer another benefit — mobility. They're usually light and can often be easily carried to a work site.

Saws — Tools for Cutting. An adequate collection of handsaws should include a crosscut saw for cutting materials to size, a backsaw with miter box for angles, a coping saw for trim and finish work, and a keyhole saw for cutouts.

Factors involved in selecting saws include blade size and the number of teeth along the blade. The term *point* indicates the number of teeth per inch and the tooth size. Generally, the more teeth, the smoother and slower the cut. The term *set* refers to the distance from which the teeth project from the side of the saw blade. This slight distance on either side of the blade determines the width of the cut or *kerf.* A smaller set produces a finer kerf.

Choose a crosscut saw made of high-quality steel for your workshop. Make sure that the handle feels comfortable and check the back of the saw for trueness. The best saws are *taper-ground,* which means that the blade is slightly tapered in two directions — toward the back and toward the tip. This feature helps to prevent the saw from binding.

The backsaw, so named for its rigid back, has a thin blade which allows for very fine cuts. Teeth per inch can range from 10 to 16; choose a 12 *tpi* type for carpentry work and one with more teeth for cabinetmaking. A miter box which accompanies the backsaw positions it for 90°, 60°, 45° and 30° angle cuts.

Coping saws, excellent for cutting tight curves, are somewhat limited by the size of their frames; common throat depth is 4¾ inches. Blade length runs approximately 6½ inches with 12-15 teeth per inch as the norm. The coping saw is versatile; its blades may be positioned in any direction for pushing or pulling the saw, or the blade may be removed and then refastened and used through a pilot hole.

The keyhole saw also has a thin blade; it too is used for cutting curves. Primarily it is used for starting cutouts, often from within a drilled hole. Keyhole saws sometimes come in sets with several types of interchangeable blades. A good all-purpose blade measures about 10 to 14 inches long and has 8 teeth per inch.

Hammers and Mallets — Tools for Fastening. Depending on what your projects are to be, you'll need to choose hammers and mallets to match. Some factors in your choice are your own strength, the type of materials you'll be working with, and the cost.

Claw hammers are the standard. A *ripping-claw* model has a slightly straighter claw than the more common *curved-claw* hammer, the one used to easily pull out nails. Hammers are sized by their head weights, ranging from 7 to 28 ounces. For cabinetmaking choose whatever feels comfortable to you. For carpentry choose a fairly heavy model that will help with the work. Your arm can tire quickly with a lightweight hammer because of all the arm strength needed to drive nails.

Another thing to consider is the face, the part that strikes the fasteners. It may be flat, slightly convex, or meshed. The flat and meshed types strike more surely, but they will mar your work surface. The convex type, used properly, will not dent the surface but tends to glance off nail heads.

For lightweight fasteners, such as tacks or brads, equip your shop with a tack hammer. Another cousin to the hammer, the mallet, is a workshop essential. Mallets are available in a range of weights and are composed of several different materials, such as rawhide, rubber, plastic or wood. Keep several types of mallets handy for such chores as knocking members together, driving in dowels and working with hand chisels.

Chisels and Planes — Implements for Shaping. Often workpieces require shaping and paring before members can be assembled or installed, and this is when chisels and planes are put to use. A chisel may be used to notch framing members or to make mortises for door hinges. There are many different kinds of chisels available and they are often sold in sets. Purchase a basic set of *bench chisels* and add other types as you need them.

Two types of planes are suggested for the beginner. The block plane which is relatively small, is for smoothing end grain, cutting bevels and trimming off excess stock. The larger smooth plane is for small to medium-size jobs. However, if you want a plane that is made specifically for large rough surfaces, choose a jack plane.

Tools for Drilling and Fastening. A good power drill is relatively inexpensive and it has so many capabilities that practically everyone owns one. Still, there is cause for having drills operated by muscle alone.

A brace and bit can be used for tough jobs that would overtax the average ⅜-inch electric model. The best kind of brace to acquire is the type that

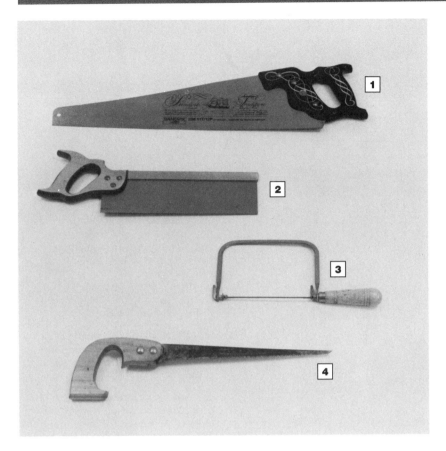

Saws

1 **Crosscut Saw.** A standard saw for cutting across wood grain; may also be used to cut sheet materials. Varies in lengths between 20 to 26 inches long and has 4 to 12 teeth per inch.

2 **Backsaw.** A saw with a strong back and fine teeth. Usually comes in lengths between 10 and 20 inches. Most often used along with a miter box for cutting trim.

3 **Coping Saw.** A saw with a thin, fine-toothed blade. Normally used to cut curves and odd shapes in plywood, hardboard and paneling.

4 **Keyhole Saw.** This saw with a fine-toothed tapered blade enables you to do fine work and cut small openings.

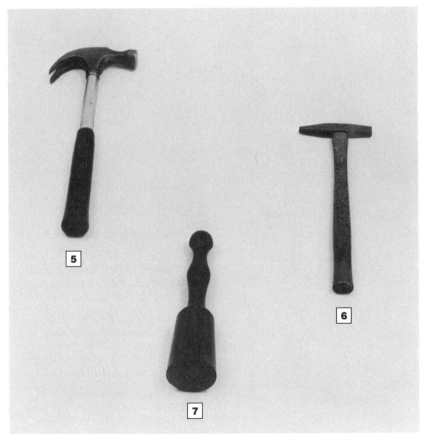

Hammers

5 **Claw Hammer.** Used to drive and pull nails. Usually made with a sturdy wooden handle and metal head, the claw hammer should be selected by weight.

6 **Tack Hammer.** Especially useful in cabinetmaking for driving brads or tacks. An option is to use an 8-ounce claw hammer.

7 **Mallet.** This multi-purpose tool, made of wood, rawhide, plastic, or rubber, is used to tap in dowels, knock stubborn joints, and drive other tools. Weights vary.

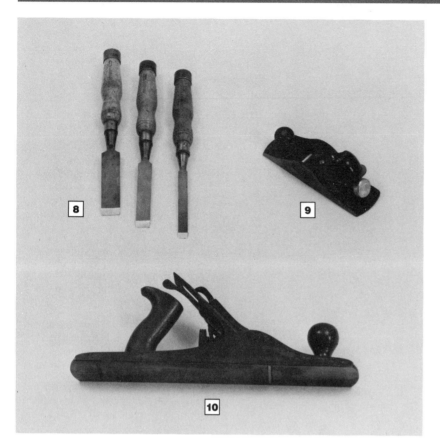

Chisels and Planes

8 **Chisels.** Used to cut mortises, notches and grooves; chisels vary considerably in style. Some have beveled edges, others have square edges. Handles may have steel caps for driving with a hammer. Sets of chisels usually include blade widths from ⅛ inch to 2 inches.

9 **Block Plane.** A small plane, typically 6" by 1⅝" that is used for smoothing end grain, trimming, and cutting bevels.

10 **Smooth Plane.** A general purpose plane measuring from 7 to 9 inches. (An option for large rough surfaces is to use a larger jack plane.)

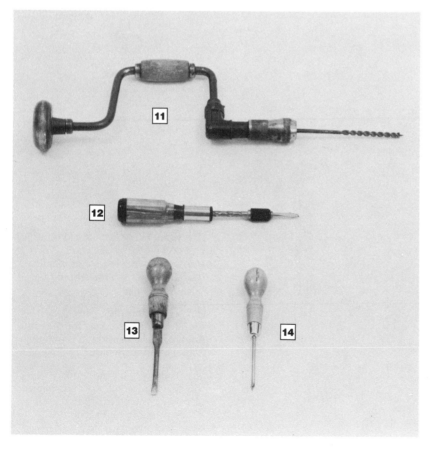

Drills and Screwdrivers

11 **Brace.** Operates in a swing motion or by a ratchet; accepts many types of bits. A popular size is the 10-inch model.

12 **Yankee Drill.** Pushing on the handle turns the drill's mechanism. Used to drill small pilot holes.

13 **Blade Screwdriver.** This type is often referred to as the standard screwdriver. It comes in numerous sizes and is used to remove and drive screws that have slotted heads.

14 **Phillips Screwdriver.** This type comes in several sizes and is used to remove and drive Phillips head screws.

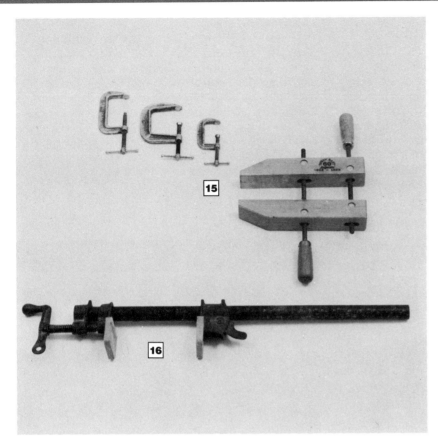

Clamps

15 **C-Clamp and Handscrew Clamp.** They are used to hold glued stock together or secure workpieces to a table or workbench. Both types are available in a wide variety of shapes and sizes. Because they are made out of wood, hand-screw clamps are less likely to mar the stock.

16 **Bar Clamp.** Used for edge-to-edge gluing. Made of wood or metal; available in a variety of sizes.

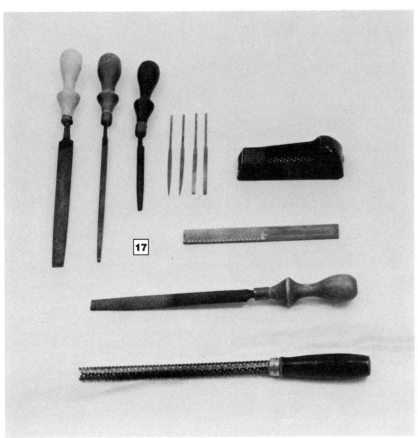

Hand Tools for Smoothing

17 **Files and Rasps.** Files and rasps come in many different styles and shapes. Generally, files may be used on both metal and wood; rasps, identified by their triangular teeth, are to be used on wood only.

includes a ratchet. The ratchet allows you to drill when there isn't sufficient space to swing a brace handle a full 360°.

There are times when electricity is not readily available or the weight, bulk, and power of an electric drill are too much to handle, such as when you want to drill small pilot holes in cabinetry. A Yankee drill comes in handy for these tasks.

For installing and removing screws, you'll need an assortment of standard and Phillips screwdrivers.

Clamps — Tools for Holding. No matter how many sets and types of clamps you accumulate, you just never seem to have enough. They are used to hold glued stock together, to hold workpieces together for fastening, and to secure stock to a table or bench. Probably the most familiar are the C-clamp and handscrew clamp. Bar clamps have one fixed and one movable jaw; they are normally used to hold wide glued-up stock.

Files and Rasps — Tools for Smoothing. Keep a selection of files and rasps in your workshop for easy hand-smoothing. Some common shapes are flat, half-round, square and round. Rasps are used exclusively on wood but files may be used on either wood or metal. Both are identified by their tooth pattern, tooth coarseness, length, and shape. Use a *single-cut file* for precise work and a *double-cut file* for speed. The latter has a criss-cross pattern. All rasps have triangular teeth, so they won't clog as easily as files.

PORTABLE POWER TOOLS AND ACCESSORIES

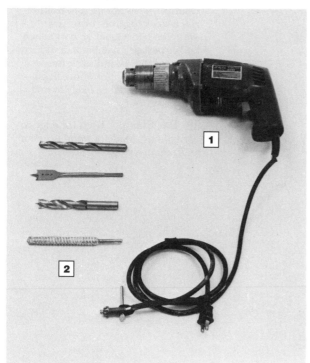

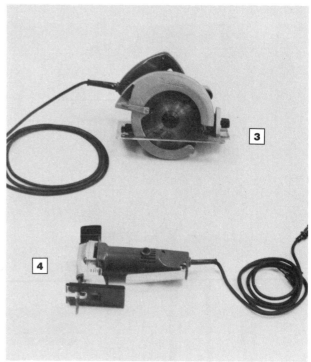

Drill and Drill Bits

1 **Power Drill.** Choose a ⅜-inch power drill for both cabinetmaking and carpentry tasks. Recommended features are variable speed and reversing capability.

2 **Common Bits.** Shown top to bottom: twist, spade, brad point, and masonry bits.

Saws

3 **Circular Saw.** Primarily used for straight cutting of lumber and plywood. A 7¼-inch model is suitable for most projects. The saw should have depth and angle adjustment levers, a fixed blade guard and a lower spring-action blade guard. A ripping fence, available on some models, helps guide straight cuts.

4 **Sabre Saw.** Operates in an up-and-down motion for cutting curves and bevels. Available in single-speed, two-speed, and variable-speed models; choose the latter for fine work and more control.

Portable Power Tools for Speed

By investing in portable power tools, you'll save valuable time and effort otherwise spent doing everything by hand. Often, too, you'll see finer craftsmanship, since these tools are engineered for balance and alignment. The tools that follow will be excellent investments for your workshop but there will be yet another payoff. These tools are portable; they can be carried to a work site.

Cordless models of the following tools are also available. This is a real plus since they can be used at work sites where electricity is not available. One thing you'll need to remember, however, is to keep them charged or they won't be ready when you need them.

The Versatile Electric Drill. Electric drills are classified by their torque and by the maximum size bit shank that their jaws, or *chucks* will accept. The larger the chuck, the greater the torque. However, a higher torque has a negative effect on the speed of the drill. Therefore, where medium-range power, speed, and chuck size are needed, you should choose a ⅜-inch type.

Drills are available in single-speed, two-speed, or variable-speed models. Another optional feature is reversibility. The variable-speed reversible drill is recommended since it is the easiest to handle when starting holes, drilling metals, or driving and removing screws.

Bits are available for a variety of uses. But, at least in the beginning, you should use your drill for its

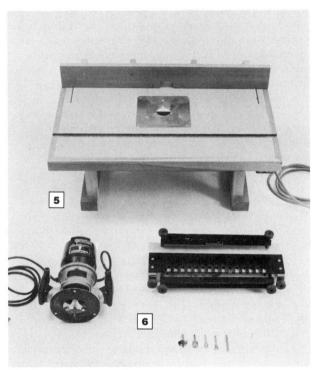

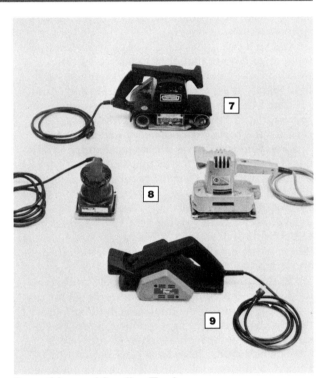

Router and Accessories

5 Router. Used for making grooves, mortises and other joinery, rounding off edges, trimming laminates, and cutting decorative internal and external edges. Select a router with at least 1 horsepower.

6 Basic Accessories. Router bits come in a variety of shapes and sizes, each designed to perform a specific operation. A *router table* accommodates the router in an upside-down position, allowing you to move the workpiece instead of the tool. The *dovetail jig* aids you in making fine joints such as those used in drawer-making.

Smoothing and Sanding Tools

7 Belt Sander. This sander is useful for abrading rough surfaces quickly. A popular all-purpose size is 3 inches by 24 inches.

8 Finishing Sanders. Two types are available: The *orbital sander* (left) moves sandpaper in a circular motion, whereas the *straight-line* type (right) moves sandpaper in an up-and-down motion. Some models combine both features.

9 Power Plane. A portable power tool that allows you to plane wood with considerably less effort than you would with a hand plane.

primary purpose, boring holes. The most recognizable type bit is the *twist bit* which is used to drill holes in wood and metal. *Spade bits* are used for drilling large holes in wood, drywall and plastic. *Brad point bits* should be used when appearance is important. They feature a center point that aids in locating the hole and spurs that shave the sides of the hole clean. *Masonry bits* are especially designed to withstand wear when used to drill concrete or masonry.

Circular Saw — Workhorse for the Carpenter. If you plan to set up a workshop and you have the space, you'll probably want to invest in a stationary table saw or radial arm saw. But if you'll be working outdoors, the hand-held circular saw is for you. This tool permits great portability; all you need are a few sawhorses and a power outlet and you're on your way to framing. A circular saw will allow you to cut at least ten times faster than with a crosscut saw.

When shopping for a circular saw, make sure that you get these features: depth and angle adjustment levers, an upper fixed blade guard, and a lower spring-action blade guard. Also, a helpful accessory to consider is a *ripping fence*.

Sabre Saw. This portable saw which operates in an up-and-down motion is capable of cutting wood, plywood, hardboard, laminates, plastics and light metals. It is great for cutting curves and cutouts.

The Multi-Purpose Router. The router is a 'power shaper' with amazing versatility. It can be used to cut grooves, dadoes and dovetails; and decorate and round over the external and internal edges of workpieces. By attaching special bits, you can use a router to create hinge mortises and to trim plastic laminate.

Because you will make high demands of this tool, it pays to get a model with at least 1 horsepower. Other features to shop for are: a trigger switch located in one of the handles (instead of at the top or side) and a depth-of-cut adjusting ring.

Efficient Electric Sanders. Since sanding is one of the most tedious of all woodworking chores, and since it is one of the final steps in the process, there is a great desire to get it done quickly. Electric sanders get you through the chore at top speed.

The belt sander has a hard platen upon which a 'ring' of sandpaper is fitted. It works at high speed to quickly abrade surfaces. Mostly used for rough sanding, it can also be used for trimming and fitting assembled joints.

Finish sanders work either in a straight line, in an orbital motion, or in a combination of both motions to smooth surfaces. They come in various shapes and styles but all have a soft, sponge-like material for backing up the sandpaper. For this reason they are also referred to as pad sanders. Some things to consider when purchasing a finish sander are comfort, weight, and the size of sandpaper sheets that the tool will use.

Power Planer. One fairly uncommon portable power tool that's especially valuable is the power plane. This tool makes your stock preparation and assembly tasks much easier. A good model will easily

remove ¼ inch thickness from a board with little effort on your part. The same model will produce paper-thin shavings, a feature that allows you to carefully plane surfaces without making the costly error of cutting too deep. This is especially important when you're trimming a door or other large structural component.

Power Tools That Stay Put

Stationary power tools require a substantial investment and usually call for a workshop or extended workspace. If you're interested in a fully equipped shop, you should consider buying a table saw, radial arm saw, bandsaw, jointer, planer, drill press, belt sander and disc sander. Another option is to look at 'combination' tools — several tools in one housing with a common power source. All stationary power tools will pay for themselves in precision, speed, and ease of use. In most cases, your work will be well supported and at eye level, important factors when you're working for extended periods of time.

On the contrary, you won't need a workshop full of large tools to build a simple cabinet or make a room addition. The two stationary power tools that are recommended as *basic* to any shop are the table saw and the radial arm saw. (You should have one or the other.) Aside from those two, the next ones in importance, and ones you should consider purchasing are the jointer, planer, belt sander and disc sander. All these are described below. **Be sure to follow the manufacturer's instructions for proper operation of these tools.**

The Table Saw. The table saw, as its name implies, incorporates a work table as part of the tool. This work surface not only supports your work, but accommodates a miter gauge, rip fence, and other jigs and fixtures for precision cuts.

The saw's *rip fence* slides across the table parallel to the blade; its function is to guide stock that is being ripped. The *miter gauge* serves to hold stock that is being crosscut. The table or the blade can be tilted for cutting angles. All of these capabilities make the table saw an extremely versatile tool.

Table saws are sized according to the diameter of the saw blade; the 10-inch size is recommended for the home workshop. With it, you can make cuts up to 3¼ inches deep.

The Radial Arm Saw. The radial arm saw combines the flexibility of a circular saw with the precision and ease common to all stationary power tools. In the past, it was primarily used by carpenters to crosscut long stock. However, the modern tool has a remarkable amount of auxiliary functions. With special jigs and accessories, it will do drilling, sanding, shaping, sabre sawing, and routing. It is an excellent tool for the woodworker with limited space who wants to move up from portable power tools.

State-of-the-art radial arm saws feature electronic programmable settings for blade height and speed — to make your work even easier. Like the table saw, the tool is sized by its blade. A 9- or 10-inch model should suffice for the home workshop.

The Jointer and Planer. These two power tools are sometimes confused with each other but they do perform separate, distinct functions. When you use rough-cut lumber, you need to machine it to a uniform thickness. The tool for this process is the planer, or *thickness planer,* as it is sometimes called.

Another frequent problem with lumber is rough or irregular edges. The tool for this problem is the jointer. The jointer squares up the edges and ends of boards so that they will be perfectly sized to join with

The table saw is an excellent tool for ripping wood and making special cuts quickly and accurately. With the help of accessories such as dado blades and molder knives, it will make grooves and other joinery. Though models vary, they all have these characteristics: a table that supports the stock during the cut, and a saw blade which mounts to an arbor and projects through a slot in the table. Some table saws have a rip fence and miter gauge that help support and guide the stock during the cut. On most models, the arbor also tilts, allowing for even more flexibility.

The radial arm saw is like a portable circular saw mounted on a large, strong arm. It offers the advantage of pushing the work to the blade, instead of vice-versa. Additionally, the tool may be raised, lowered, tilted, or swung into various positions. New models include accessories for drilling, sanding, and shaping.

A jointer is used primarily to square the edges and surfaces of boards. Straight square stock results in tight fitting joints, easy to assemble components and professional looking projects. Often used in conjunction with a table saw before and/or after sawing.

Use a planer to smooth the surfaces of rough lumber, even out the surfaces of glued-up stock, and reduce any number of boards to the exact same thickness.

THE RIGHT TOOLS

other boards. So in some cases the jointer is used before the planer and in other cases, after.

The jointer can be used to prepare wood for your projects, but it also can be employed *during* construction — for smoothing faces and edges, and for cutting rabbets, tapers, and bevels. Although the tool is expensive, it is a precision machine that can save you trouble 'down the road'. Jointers are sized by the maximum width of stock that they'll accept and also the maximum depth of cut. If you intend to do a lot of woodworking, consider getting a 6-inch model with a ½-inch depth of cut. If you're not planning to use the jointer to a great extent, a smaller model would be sufficient.

The planer is used to surface rough lumber to a specific thickness. Obviously, it can save you hours of hand-planing. Some planers also have the capability of creating moldings. The planer is sized by the width of stock it accepts. A capable tool will plane boards measuring up to 12 inches in width.

The Belt Sander and Disc Sander.   These sanders allow you to present the work to the tool instead of vice-versa. The belt sander works well for sanding the end, edge and surface of wide and thick stock, while the disc sander works best for smaller stock. Both sanders are sold in standard sizes so that you can easily purchase abrasive discs and belts to fit them. Discs are self-adhesive or Velcro®-attached for easy installation.

Disc sizes vary; they're available in 6, 9, and 12 inches. The 12-inch model is considered standard. Belt sizes vary also; two popular sizes are the 4" x 36" and the 6" x 48" belts. (The two larger dimensions indicate the full length of the abrasive belt.)

These machines have more than one speed setting. Accordingly, you can set the machine for the particular material that you're sanding and also the type of abrasive you're using.

Strictly for Support

The *table extension* is an accessory for the table saw — or it may be an integral part of the tool. Basically table extensions are projecting metal arms that are used to hold large and unwieldy work such as sheet materials.

If your table saw doesn't have extensions or if the extensions are not long enough, then you might consider investing in a *roller stand*. This will act as a support when you're working with unusually long stock or large pieces of sheet materials. You might want to consider two roller stands which would allow you to support the stock at both the infeed and outfeed ends of the table saw. Roller stands are also used to support long stock on other tools such as the planer and jointer.

Sawhorses.   These are the woodworker's 'trusty friends'. Almost always used in pairs, they can be moved around easily to support all sizes of stock. In selecting your sawhorses, you'll find many options. You may purchase all-metal ones or the traditional wooden types. Also popular are the metal brackets that hold sawhorse legs. If you choose these, then you need only to make the legs and crossbraces yourself out of 2 x 4 lumber. An advantage of this type is easy storage; they may be broken part since the legs are not held by screws. Yet another product is metal legs to which you add only the crossbraces.

The belt sander and disc sander are excellent sanding tools. The belt sander (on the left) sands the surface and edges of stock with ease, while the disc sander (on the right) works best for sanding the end grain.

On page 22 are detailed plans for making a unique type of sawhorse that will give your stock excellent support during various woodworking operations. A special feature allows you to rip and crosscut lumber more easily.

Whether you choose to buy or make your own, keep your comfort in mind. If you're shorter or taller than the norm, build sawhorses to a complementary height. Start with the average height of 20 inches and vary the dimension to suit your needs.

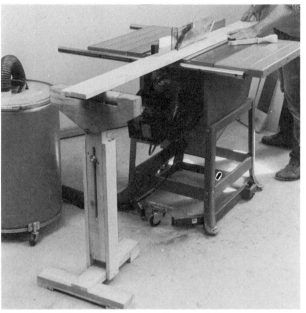

Support for Large Stock. When crosscutting or ripping long stock and sheet materials on a table saw, you will need extra support. Some table saws are equipped with exten- sion tables. If yours is not, use roller stand(s) to support the stock as it exits the saw.

Using a Special Sawhorse. This sawhorse, which you can construct yourself, is made from common stock. The top is made from two 2 x 4s, instead of the usual one. These are spaced to create a gap. The user simply lays boards across the gap and positions the blade over the gap for easy cutting.

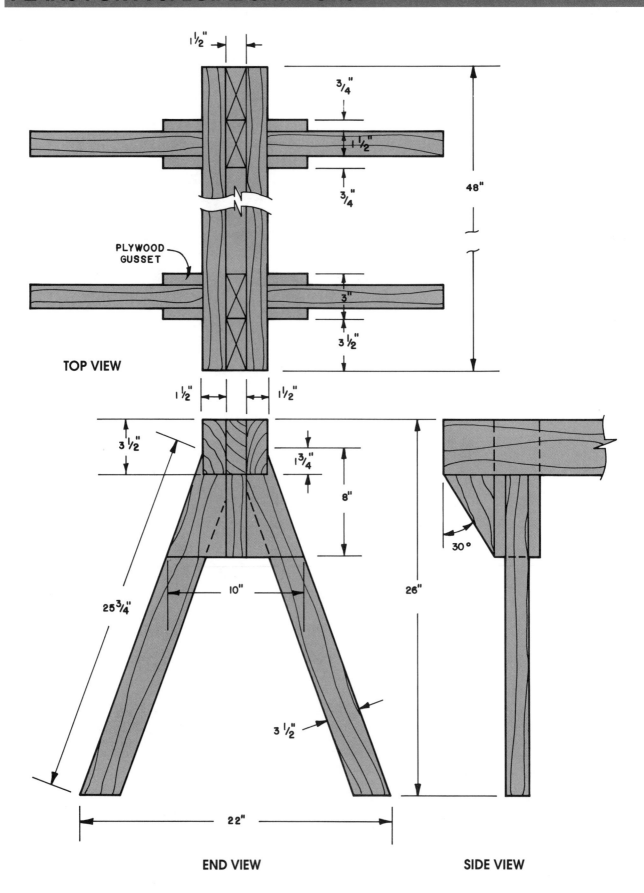

TOP VIEW

PLYWOOD GUSSET

END VIEW

SIDE VIEW

JOINERY DETAIL

BILL OF MATERIALS
FINISHED DIMENSIONS IN INCHES

A. Legs (4)   1½ x 3½ x 25¾
B. End braces (2)   1½ x 3½ x 9¾
C. End spacers (2)   1½ x 3 x 9¾
D. Gussets (2)   ¾ x 8 x 10
E. Horizontal   1½ x 3½ x 48
    members (2)

3

Woodworking Materials and Hardware

The main material you'll use when adding on or building in is wood. So before you go to the lumberyard you should know what type of wood you'll need, the grade and the amount. This chapter will tell you the basic information you'll need to know about selecting wood and wood products (plywood, particleboard, hardboard) for your projects.

If you'll be building an add-on, your use of materials will be limited to only a few kinds of wood and wood products, mostly softwoods. But for built-ins, the selection is broader; you can use both softwoods and hardwoods. When the appearance of the finished project is important, you might be tempted to spend a lot of money on fine hardwood. Therefore, it's recommended that you go easy especially if this is your first woodworking experience. Good grades of softwoods are excellent for making built-ins and are more economical.

Also covered in this chapter are the additional materials you'll need. Glues and fasteners are presented along with charts to help you choose specifically what you need. Cabinetry hardware is shown, to give you an idea of available styles and to spur on your creative thinking.

With the information in this chapter, you'll be able to make out your shopping list and choose the proper materials for your project.

Timber Talk — A Short Lesson on Woods

Wood is divided into two main categories: *softwoods* and *hardwoods*. These terms do not necessarily describe the wood, however. Instead they refer to the trees from which the wood is derived; softwoods come from conifers and hardwoods from deciduous trees. Solid cut pieces of these woods, called *boards*, *planks*, or *timbers*, are all referred to as *lumber or stock*. Other wood products are not solid but are plied, glued or compressed together. They include *plywood*, *particleboard*, and *hardboard*.

Woods and wood products vary in how they are cut; their strength; and how they are sized, graded, and priced. In choosing stock, a good rule of thumb is to use the quality that will serve the purpose and is acceptable to your tastes. Cabinet backs, drawer bottoms, and the backs of facing pieces are all places where you can use stock with cosmetic defects or of lesser grades.

When a fine appearance is crucial to the finished look, choose unblemished stock, but understand that you will pay a premium price for it. Or, where stability is crucial, as for door framing where closet bifold doors will be fitted, select 'clear' stock; knots or warpage could impair smooth operation of the doors. Also, make your selection based on the final finish.

For a natural-stained or clear finish, choose better grades of wood and plywood; for a project that is to be painted, choose more economical grades. When buying plywood, it's extremely important to think of the final finish. Plywood comes plain, one side veneered, or both sides veneered.

Softwoods. The most commonly used kinds of lumber for home improvement projects are softwoods — the most popular and widely available of these being white pine, yellow pine, cedar, and redwood. Terminology might vary in different localities but generally softwoods are divided into three categories for the purpose of grading. The first two categories are *clear* and *common lumber* — both with a nominal measurement of 1 inch thick. The third is *dimension lumber* — lumber with a nominal thickness of 2 inches.

Clear lumber is more expensive, has fewer blemishes and is rated A through D. Common lumber has more knots, is less expensive and is rated No. 1 through No. 5. Dimension lumber, mostly used for framing, but sometimes used for cabinets or smaller projects, is available in four categories: construction, standard, utility, and economy. Use the chart on this page to determine which grade of softwood to select for your particular project.

Softwood pieces are usually surfaced on all four sides (S4S) at the lumber mill and the ends will reveal

SOFTWOOD LUMBER GRADING

CLEAR LUMBER		COMMON LUMBER		DIMENSION LUMBER	
(nominal thickness 1 inch)		(nominal thickness 1 inch)		(nominal thickness 2 inches)	
Grade	Description	Grade	Description	Grade	Description
A	No blemishes. Ideal for natural finishes.	No. 1	Large, tight knots. Acceptable for painting.	Construction	For use where sturdy framing is needed.
B	Tiny blemishes. Fine for natural finishes.	No. 2	More knots; some loose. Can be painted.	Standard	Acceptable for framing.
C	Some small blemishes and tight knots. Fine for natural finishes and paint.	No. 3	Splits and holes. Unacceptable for painting.	Utility	Not for framing; suitable for cabinets and other small projects.
D	Several small knots. Acceptable for painting.	No. 4	Many splits and holes. Some sections can be salvaged. Do not paint.	Economy	Avoid this grade.
		No. 5	Unsuitable for cabinetry projects.		

that they are sawn in one of two ways: flat grain or vertical grain. There might be a price difference, but choosing the vertical sawn grain will help to ensure against warpage.

Look for the usual lumber defects — knots, separations, pitch pockets and warpage — and decide how you will handle them before taking your wood home. Many minor problems can be remedied or, if your project layout permits, you can salvage what you need and scrap the sections with the worst defects. Determining whether lumber is correctly dried or still 'green' is difficult, so to be safe, always buy kiln-dried *(Kd)* lumber rather than the less-expensive air-dried *(Ad)* variety.

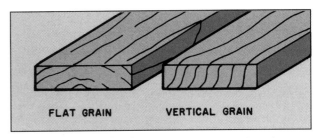

How Lumber is Milled. There are two ways boards are cut from logs at the mill: flat grain and vertical grain. Flat grain sawn boards have a grain that runs the width of the cut, vertical grain sawn boards have a grain that runs from the top to the bottom of the board — either straight up and down or at an angle.

CHECKING FOR AND WORKING WITH LUMBER DEFECTS

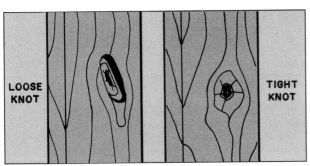

Knots. The most common lumber defect is knots. There are two basic types: *loose* and *tight*. Knots themselves are strong but the wood around knots is weakened. Loose knots fall out easily but may be glued back into place. Tight knots may be filled and sanded lightly for painting over. Knots do not always detract from a piece; some woodworkers prefer to leave them visible.

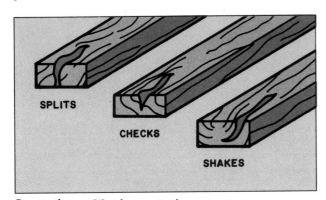

Separations. Wood separates in three different ways resulting in *splits, checks,* or *shakes*. Splits are cracks that extend all the way through the wood, checks are cracks along the growth rings, and shakes are hollows between growth rings. If you cannot cut away a section of board with these defects, an option is to fill them with patching compound.

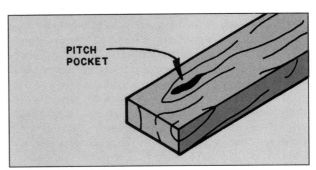

Pitch Pockets. Found in pine and other evergreen woods, these are openings that ooze resinous pitch. Difficult to fill, the pockets will usually bleed. If you must use wood that contains this defect, first clean away the pitch with turpentine and then apply shellac to the surface before adding the final finish.

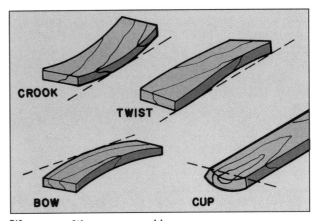

Warpage. Warpage, caused by improper storage or drying, is not included in the grading of lumber. Therefore, boards should be scrutinized for crooks, bows, twists, and cups. To avoid these problems, always purchase kiln-dried wood.

WOODWORKING MATERIALS AND HARDWARE

Sizing is another important matter. Learning the terminology will help you when the time comes to design and lay out your project. Traditionally, in the lumber industry, *nominal sizing* refers to the depth and width of lumber that has not yet been dried and surfaced. Thus, when you purchase a 2 x 4 you will get a board that was once that size but now measures 1½ x 3½ inches.

Consult the chart on this page (right) for the conversions from nominal to actual sizes and keep it handy when you plan your projects. Larger sizes, 4 x 4 through 8 x 8, are not required for most home projects, so these are not included in the chart.

Nominal sizing does not apply to any pieces of softwood that are smaller than a 1 x 2. Such smaller pieces are considered molding and are sold in exact sizes. A board's length, also, will be the actual size. Typically, softwood comes in even-numbered lengths of 4 to 20 feet. However, the lumberyard will cut it to the length you need or to a size you can easily and safely transport. Some lumberyards will deliver, especially if you buy a large quantity of materials. So plan ahead and if possible purchase all of your materials at one place.

Hardwoods. Hardwoods are distinguished by their beautiful colors and grain patterns. Depending on the species, hardwood can be costly and is sometimes not readily available. For these reasons, woods like oak, poplar, maple, walnut, cherry, and mahogany might only be available in random lengths, limited sizes, or in veneers.

Local lumberyards do not usually carry hardwoods, with the possible exception of oak and poplar. Profesional cabinetmakers deal with wholesale suppliers to obtain their hardwoods in large quantities. One option for you is to make contact with a cabinetmaker in your area and place orders through his business.

Generally, hardwoods come in higher grades than softwoods with the number of defects being the determining factor. The accompanying chart lists grades up to No. 2. Though other grades are available, their quality would be unsuitable for cabinets.

Lumber of the hardwood variety is usually sold surfaced on two sides (S2S) or three sides (S3S). Also, the price might vary according to the grain pattern (flat sawn or vertical sawn). Sizing of hardwoods is actual, not nominal, and standard thicknesses (in inches) are 3/16, 5/16, 7/16, 9/16, 13/16 and 1 1/16. Widths will vary from board to board and, as mentioned previously, lengths will be random.

NOMINAL AND ACTUAL SIZES OF SOFTWOOD LUMBER

Nominal	Actual (in inches)
1 x 2	¾ x 1½
1 x 3	¾ x 2½
1 x 4	¾ x 3½
1 x 6	¾ x 5½
1 x 8	¾ x 7¼
1 x 10	¾ x 9¼
1 x 12	¾ x 11¼
2 x 2	1½ x 1½
2 x 3	1½ x 2½
2 x 4	1½ x 3½
2 x 6	1½ x 5½
2 x 8	1½ x 7¼
2 x 10	1½ x 9¼
2 x 12	1½ x 11¼

Lumber Pricing. Hardwoods are sometimes sold by the pound but the most common and most current method of pricing, for both softwoods and hardwoods, is by the running foot. Still, some lumberyards adhere to the traditional method of pricing by the board foot — a unit of measurement that is one square foot by 1 inch thick. Use the following formula to calculate board feet:

$$\frac{\text{nominal thickness (in.)} \times \text{nominal width (in.)} \times \text{length (ft.)}}{12} = \text{board feet}$$

Plywood and Sheeting Materials

Plywood is lumber that has been manufactured of thin wood layers (called *veneers*) which are plied or glued together. It has many advantages over solid lumber which accounts for its wide popularity. It is resistant to warping, very strong, and comes in large sheets. Moreover, it's considerably less expensive than solid lumber. Plywood is strong in all directions

HARDWOOD LUMBER GRADING

FIRSTS AND SECONDS	SELECT	COMMON	
No splits or knots on the front or back surface. Fine for natural finishes.	No splits or knots on the front surface; small blemishes on the back. Fine for natural finishes.	No. 1	Some splits and knots on both surfaces. Suitable for painting.
		No. 2	Numerous splits and knots; acceptable for painting.

because the grain of each veneer runs perpendicular to those adjacent to it.

There are two types of plywood — interior and exterior — determined by the glue used to hold it together. *Interior* is sufficient for indoor use in any area that won't be exposed to moisture or temperature changes. *Exterior* should be used outdoors or in other areas where these conditions exist.

Plywoods are divided, like lumber, into softwoods and hardwoods — this designation being based on the condition of the face veneers only. One face may have a different grade than the other. For example: if you were to order A-C softwood plywood, you would get a good front face and a rough back face.

Examples of *softwood plywood* are Douglas fir, pine, spruce, cedar and redwood. *Hardwood plywoods*, frequently used as alternatives to expensive solid hardwoods, include birch, ash, maple, cherry and oak. Others, domestic or imported, may sometimes be special-ordered. Grading charts are shown here for both softwood and hardwood plywood.

Plywood Core Materials. As mentioned earlier, plywood is composed of a series of veneers all pressed and glued together. While this is true for softwood plywood, it is only partly true with the hardwood type. With hardwood plywood, you will actually have a choice among several types of center or core materials.

Each kind of material offers different advantages. To begin with, *veneer-core* plywood is versatile and inexpensive but the open spaces in the edges leave a lot to be desired in terms of beauty. A trick when using it is to cover the edges with molding or with veneer tape once the project is assembled.

Options, when purchasing hardwood plywood, are to purchase *particleboard-core* or *lumber-core* plywood. Edges with such solid cores will not only look better, but they will also accept fasteners better. Particleboard cores are basically a slab of wood chips glued together under pressure. Generally, lumber-core plywood is preferable; but refer to the chart shown on page 30 if you are in doubt about which kind to use.

Plywood Sizes. An advantage of using plywood is that it comes in sheets — the standard size measuring 4 x 8 feet. At some lumberyards, half or quarter sizes are available. As for thicknesses, sizing varies according to whether it has a softwood or hardwood outer veneer. Softwood plywoods are typically available in ¼-, ⅜-, ½-, ⅝- and ¾-inch thicknesses. Thicker sheets are available but difficult to find. Hardwood plywoods come in the following standard thicknesses: ⅛, 3/16, ¼, ⅜, ½, ⅝, ¾ and 1 inch.

Particleboard. Particleboard, already mentioned as a core material for plywood, is another sheeting material. Also known as *chipboard, pressboard,* or *faceboard,* it too is available in interior and exterior grades. Because it is made of wood chips and sawdust glued together under high pressure, it presents several problems that you should be aware of. It's terribly inadequate at holding fasteners; it's very

heavy and consequently difficult to work with; and it's very hard on sawing tools — dulling them in a much shorter time span than other materials.

The benefit, of course, is cost, and particleboard is an acceptable material for cabinet backing, drawer bottoms, and small sliding doors. Joints and seams may be reinforced with wood blocks. For secure fastening, nails, screws, or bolts should be used in combination with glue. If you wish to paint particleboard,

SOFTWOOD PLYWOOD GRADING*

GRADE	DESCRIPTION
N	No blemishes or knots. Fine for natural finishes.
A	Smooth surface, tiny blemishes that are neatly repaired. Acceptable for natural finishes, ideal for paint.
B	Smooth, firm surface with tight knots and small repair plugs. Suitable for painting; unsuitable for clear finishes.
C	Rough surface with unrepaired knotholes. Accepts paint poorly. Use only in areas that will not show.
D	Large knots and holes. Use only in areas where strength and appearance are not important.

*Note: Both surfaces of softwood plywood are graded. When ordering, specify the grade desired for the face and the grade for the back; not all combinations are available.

HARDWOOD PLYWOOD GRADING*

GRADE	DESCRIPTION
Premium or A	Well-matched grain and color. No blemishes. Ideal for natural finishes. Very expensive. Not widely available; often special-ordered.
Good or 1	Color and grain are fairly well matched. Ideal for natural finishes. Expensive but worth it for favored projects.
Sound or 2	Smooth, solid surface with some small, tight knots. Suitable for painting or natural finishes.
Utility or 3	Rough surface with knotholes as large as 1 inch; sometimes with small splits. Not suitable for painting.
Backing or 4	Very rough surface, with defects and large holes. Not recommended for cabinetry projects.

*Note: Both surfaces of hardwood plywood are graded. When ordering, specify the grade desired for the face and the grade for the back; not all combinations are available.

do not use a water-based paint which would cause the base material to swell. Thicknesses from ¼ to ¾ inch are available.

Hardboard.   Yet another sheeting material, hardboard is similar to particleboard but has a smooth instead of speckled appearance. Both faces of the board may be smooth or the back may have a mesh-like texture. Also an inexpensive material, hardboard has the same drawbacks as particleboard, so you should follow the same guidelines when using it. It is usually available in ⅛- and ¼-inch sheets.

Choosing Plastic for Durability

Depending on what your project is, you may have reason to use plastic materials. Though many types and brand names have come and gone, *plastic laminate* is the tried and true favorite for kitchen countertops and other surfaces that receive constant use — or a dose of abuse now and then. Wood, no matter how it is finished, simply isn't as durable.

You might consider using plastic laminate for countertops, cabinets, and shelving. The choices will be many; laminate comes in hundreds of different colors, and a variety of patterns and textures. Though somewhat expensive, you can lay it over an inexpensive plywood base.

Often homeowners shy away from using plastic laminate because they believe that only a professional can install it. While laying large countertops is tricky and requires experience, the process is not all that difficult. By beginning on smaller projects, you can learn how to handle the laminate and adhesive and then graduate to larger projects. A portable router with a special laminate trimming bit is used to finish the edges, or they can be trimmed with special hand tools. Still, if you prefer plastic laminate, but are unsure of the process, you might want to purchase countertops or plywood sheets that are prefinished with plastic laminate. The only drawback is that the length may not be adequate and the choice of colors is usually limited.

Selecting the Right Glue

Not all glues are alike and since this substance can literally make or break your project, you should have an understanding of how they differ. Factors involved

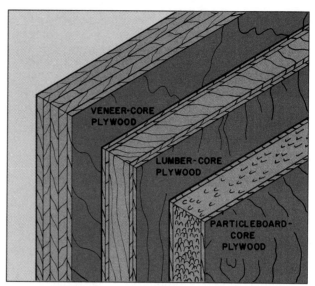

Identifying Plywood Cores.   Softwood plywood comes with only one type of core — veneer. When selecting hardwood plywood, you will have a choice of core materials — veneer, particleboard and lumber.

in making your glue selection are: the amount of strength desired, the composition of the surface areas to be glued, the preparation of the surfaces, and the fit of the joint. Another thing to consider is that some glues will stain wood or when used on natural wood some finishes will not penetrate where glue is exposed. You should take care in choosing a type that will give you the best holding power without detracting from the project's appearance.

In using glues, the proper environment must be maintained since the temperature is often critical in the bonding process. Generally, glued-up wood pieces should be clamped together, although some glues don't require clamping. Likewise, some glues require mixing while many are packaged ready to use. Use the chart on page 32 to select the proper glue for your project, but to be safe, also check the manufacturer's instructions.

Devices for Fastening Wood

When you make out your shopping list, be sure to add the fasteners; nails and screws are also very important building components.

CORE MATERIALS OF HARDWOOD PLYWOOD

CORE TYPE	THICKNESS RANGE (IN.)	DESCRIPTION
Veneer	⅛ to ¾	Good screw-holding power. Moderate cost. Exposed edges difficult to stain. Susceptible to warpage if used for doors. Difficult to saw.
Lumber	⅝ to ¾	Easy to saw. Edges are easy to trim and stain. Holds fasteners well. Most expensive.
Particleboard	⅝ to 1	Least expensive. Very stable; panels are heavy. Edges are difficult to stain. Poor edge-holding power.

Selecting Nails. Nails used in combination with glue can create strong holding power and they are quite acceptable for cabinetry projects with light and medium loads. Nails alone, or in combination with other metal fasteners, should be used for holding framing components together.

Nails are sold in lengths from 2 *penny* (or 2*d*) which measures 1 inch, to 16 penny (16d) which measures 3½ inches. All sizes in between, in ¼-inch increments, are available. A general rule for determining nail size is that two-thirds of its shank should penetrate into the piece(s) of wood that are being joined together.

There are many types of nails available, and the criteria for choosing them are: holding power, visibility, and their compatibility with construction materials. The most popular types are found in hardware stores, in 1-pound or 5-pound boxes, or sometimes loose, sold by the pound. Special varieties of nails, however, can be obtained at building supply stores. Although not all types are shown in the guide on page 33, use it as a basis for making your selection of nails for each particular job.

Selecting Wood Screws. If holding power is crucial in your cabinetry project, choose screws over nails. Another reason for this choice is that if you ever need to disassemble any parts of the project, you may do so easily.

Wood screws are available in sizes, or *gauges*, ranging from 0 to 24. The gauges most often used are No. 5 through No. 14. The thicker the gauge, the stronger the screw's holding power. The chart on page 33 shows which gauges are available in which lengths. A screw's length should be determined in the same way as a nail's length.

There are several types of wood screws; each is identified by the shape of the screw head (flat, oval or round) and the grooves in the head (slot or Phillips). Flathead screws, either slot or Phillips, are the most popular since they can be easily countersunk and concealed.

Selecting Bolts. Bolts should be used when extra holding power is needed — to hang cabinets and to secure framing members. A bolt is driven completely through wood members and secured at its other end by a nut.

There are many different shapes of bolt heads as well as differently shaped nuts. Some bolts should be used with washers. Usually, a bolt's head is driven with a wrench. Many types of bolts are produced; the three most common types are shown on page 37.

Bolts are available in diameters from ¼ to 1¼ inch, and in lengths from ¾ inch to 30 inches. When choosing bolts, select a length that will equal the sum of the thickness of the pieces of lumber plus ½ inch. Use the following as a guide for length and diameter:

THICKNESS OF LUMBER	DIAMETER OF BOLT
2 inches	¼ inch
3 inches	⅜ inch
4 inches	½ inch

Special Fasteners. There is a large selection of fasteners for special situations. Plates and anchors are used to provide extra support to framing members — at trusses and at joints. Special devices are used to attach countertops to cabinets. Some of these are shown on pages 36 and 37, but you should also scour home improvement centers and woodworking catalogs for ideas when you have a fastening task.

Hardware for Wood Projects

Whether making cabinets or building a room addition, you're likely to have a need for hardware to complement your project. Functioning types of hardware — like hinges, catches, drawer glide assemblies, and shelving standards — will directly affect the design, so you should consider these early in the planning process.

TIPS FOR BUYING WOOD AND WOOD PRODUCTS

■ Save yourself time by checking with lumberyards by phone to see if they have what you want. If you know exactly what you want, order by phone. However, if you order by phone, you won't see what you're getting until it's delivered. Then, if there's a problem, you'll waste valuable time. Until you get to know the lumberyard, it's better to pick out what you want in person.

■ Choose a reputable lumberyard that has experienced salespeople. Especially if you're a novice at checking wood and following plans, you'll need some trustworthy advice.

■ Always take your plans with you. You might need to purchase less wood than you think or you might be able to use lesser grades of wood. Often, Common grade softwood can be used in place of Clear, as long as you can cut around the sections with defects.

■ Look closely at what you're buying so you're aware of all defects. Most importantly, check boards for warpage. Caused by improper drying or storage, this problem is not included in the grading of lumber. Warpage can be detected by laying a board down flat or sighting it along its edge. If you're buying plywood, check the back of the panel for an approval stamp. Markings will include the American Plywood Association name plus the grade of the plywood (interior or exterior), and the 'face ratings'.

■ Air-dried (Ad) wood is more likely to be improperly cured. To be safe, always purchase kiln-dried (Kd) wood.

■ Check out the store's cut-up bin or shop bin for plywood scraps, especially if your project is a small one. You might be able to salvage some pieces very inexpensively that will work perfectly for your project.

Another function of hardware is to decorate. Whether your style is traditional, contemporary, or oriental, you can make your new piece blend in by incorporating decorative hardware. Knobs and pulls for doors and drawers always come to mind but don't forget that you can add additional touches such as corner trims, side trims and latch plates. Materials vary widely; everything from wood to porcelain, brass, aluminum, steel, or plastic is available.

An Array of Hinges. Hinges come in many forms and they can be quite intimidating if you're a beginner. But if you start with the simplest kind, the *butt hinge,* and understand how it works, you'll see that most other types are basically a variation of it.

The butt hinge is made of two rectangular leaves which rotate on a central pin. The two basic tech-niques for mounting it are *surface mounting* and *mortising.* Surface mounting is basically attaching the hinge to the outside of the cabinet where it will be exposed. Mortising involves making a shallow groove from which to mount the hinge so that only part of it, the pin portion, is visible.

Butt hinges can be used only on overlapping and flush doors. For other types of doors, specially designed hinges are required. Use the hinge chart on page 35 to determine which kind will be suitable for your project. Then you can begin to shop at a home improvement center or by catalog, for hinges. You'll find dozens of choices in many different metals and finishes.

Catches, Latches, and Pulls. Doors may need to be held in position, and for this purpose, you

GLUES FOR WOODWORKING

TYPE	PROPERTIES	SUITABLE FOR	LIMITATIONS
Aliphatic Resin (Yellow Glue)	Ready-to-use. Strong, tough, durable. Good heat resistance. Non-staining, bonds quickly. One and one half hours clamping.	Gluing wood to wood and plywood, wood veneers, plastic laminates to wood, leather to wood.	Lacks water resistance.
Contact Cement	Ready-to-use. Water-resistant. Adheres on contact. No clamping.	Wood veneering, plastic laminates to wood, leather to wood, rubber to wood.	Pieces cannot be shifted once contact is made.
Epoxy Cement	Must be mixed. Bonds almost everything. Water-proof. No clamping.	Gluing metal to wood and in combination with tile, glass, metal, and wood.	Not good for gluing wood to wood in large projects.
Hot-Melt Glue	Must be heated. Fast-setting. Waterproof. Flexible. No clamping.	Mass production.	Not shock resistant. Will not take stain. Some cannot stand high temperatures.
Liquid Hide	Ready-to-use. Reliable, very strong. Resists heat and mold. Three hours clamping.	Gluing wood to wood and plywood, wood veneering, plastic laminates to wood, and leather to wood.	Not waterproof, must wait before clamping.
Polyvinyl Resin (White Glue)	Ready-to-use. Sets fast. Dries clear. Non-staining. One and one half hours clamping.	Gluing wood to wood and plywood, wood veneering, plastic laminates to wood, leather to wood.	Not good for high stress; lacks water resistance. Under heat, it will soften.
Powdered Casein	Must be mixed. Strong, fairly water resistant. Three hours clamping.	Gluing wood to wood and plywood, wood veneering, plastic laminates to wood.	Not good for outdoor uses. Will stain some woods.
Resorcinol Resin	Must be mixed. Fully waterproof; very strong. Sixteen hours clamping.	Gluing wood to wood and plywood, wood veneering, plastic laminates to wood.	Good for outdoor uses. Dries to a dark coat and glue line.

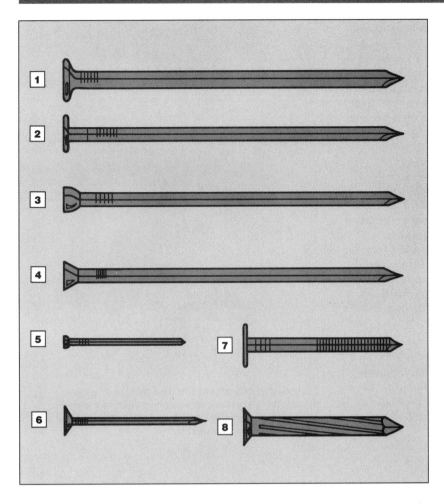

1 **Common Nail.** Great holding power because of its large head, but should be used only where it won't be seen.

2 **Box Nail.** Similar to the common nail, but with a slimmer shank. Easily bent, but less likely to split the wood.

3 **Finishing Nail.** The head is very small, permitting it to be countersunk and concealed, thus creating finished appearance.

4 **Casing Nail.** A finishing nail, but with a thicker shank and more angular head; used for setting flooring and adding casings around doors.

5 **Brad.** A small finishing nail; used on small projects and for attaching moldings to cabinets.

6 **Drywall Nail.** Coated with cement, with a special head for recessing below drywall surface.

7 **Roofing Nail.** Made with a wide head and usually galvanized. Purchase the type recommended for your roofing material.

8 **Masonry Nail.** Case-hardened, with a fluted shank; for use in fastening sole plates to concrete or furring strips to masonry.

WOOD SCREWS — GAUGES AND LENGTHS

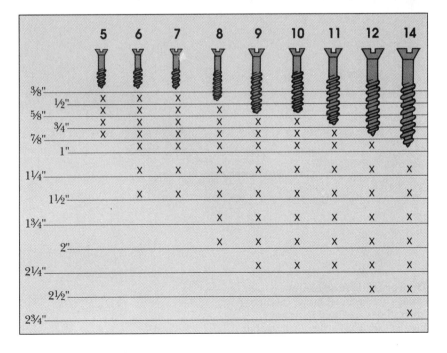

	5	6	7	8	9	10	11	12	14
3/8"									
1/2"	X	X	X						
5/8"	X	X	X	X					
3/4"	X	X	X	X	X	X			
7/8"	X	X	X	X	X	X	X		
1"		X	X	X	X	X	X	X	
1 1/4"		X	X	X	X	X	X	X	X
1 1/2"		X	X	X	X	X	X	X	X
1 3/4"			X	X	X	X	X	X	X
2"				X	X	X	X	X	X
2 1/4"					X	X	X	X	X
2 1/2"								X	X
2 3/4"									X

Shown in this chart are typical size (gauge) wood screws. Screw lengths are indicated at the left and for every X shown under a screw, that particular gauge is available in the given length. For example, a No. 7 gauge screw is available in lengths of 3/8, 1/2, 5/8, 3/4, 7/8, 1, 1 1/4, and 1 1/2 inches, as is a No. 6 gauge screw.

When selecting the length, consider what you're fastening together and measure so that two-thirds of the shank will penetrate into the adjoined piece(s) of wood.

Though only flathead slot screws are shown here, other types such as round, oval and Phillips head are also available.

should use catches and latches. These come in a wide variety of styles and materials; the most popular ones are shown on page 35.

Pulls, of course, provide a way of opening doors, and there are literally hundreds of styles to choose from. There are rules for adding pulls or knobs, so be sure that you're choosing the correct size and the correct amount of hardware as you plan your project.

Miscellaneous Hardware. Familiarize yourself with other types of hardware. Shelving projects become much simpler with the use of metal shelving standards and brackets. Likewise, drawer glide assemblies make for easy installation of cabinet drawers.

Only the very basic types of hardware are presented here. If you do a little research, you'll find an abundance of specialty types available. There are low-profile style catches, push-latches that automatically open a door and spring-loaded supports. Depending on what your cabinet will be used for, you will want to take advantage of these well-engineered devices. Shop for these items at home improvement centers, or, for an even greater selection, check specialty woodworking and hardware catalogs.

Other Materials

If you're planning to do your own masonry, plumbing, and wiring, you'll need to refer to books from your local library or bookstore to learn about the materials needed for these tasks. Then add these materials to your shopping list.

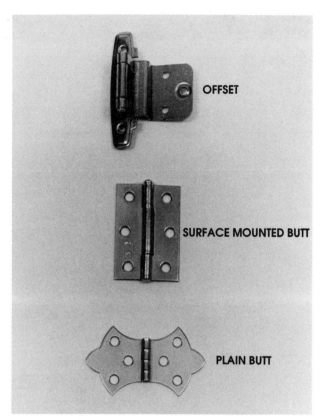

Hinges directly affect the design of your project. Several basic types are shown here: *Plain butt, surface-mounted butt,* and *offset. Piano, semi-concealed, concealed,* and *knife hinges* are also available.

MISCELLANEOUS HARDWARE

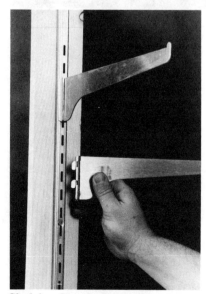

Shelving Standards and Brackets. Metal standards are installed inside cabinets; brackets are positioned at varying heights to support shelves.

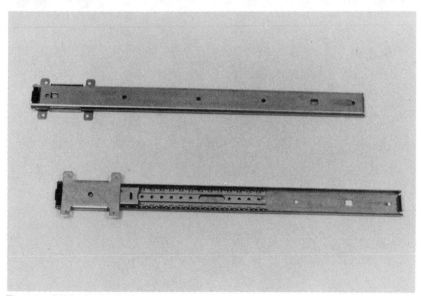

Drawer Glide Assemblies. Outer casings are installed inside cabinetry; inner glides are fitted to drawers. Easy to use; many styles are available.

CATCHES AND LATCHES

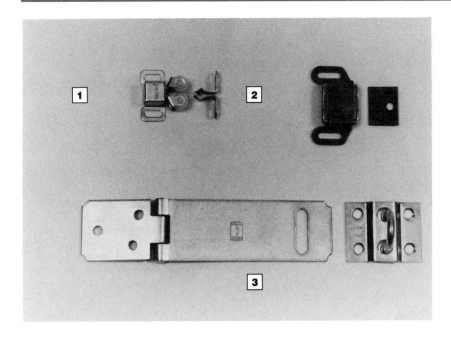

Use catches and latches to secure doors. Shown top to bottom:

1 **Friction Catch.** Consists of an arrow-shaped 'strike' which is mounted on the inside of the door and a spring-loaded receptacle mounted inside the cabinet. Installation requires precise measuring.

2 **Magnetic Catch.** Composed of a magnetic strike plate mounted inside the cabinet and a matching metal plate inside the cabinet door. More popular than the friction type.

3 **Hasp.** Used primarily in carpentry, the hasp includes a 'staple' and a slotted, hinged plate that fits over it. A pin or lock is inserted to secure the door.

HINGES AND HOW THEY WORK

TYPE	DESCRIPTION	USED FOR
Plain Butt	Size ranges from very small to very large. Consists of two leaves held in place by a removable pin.	Small projects, cabinet doors, and house doors. Surface mounted or mortised.
Surface-Mounted Butt	Like the plain butt, except usually very decorative.	Same as plain except surface mounted.
Piano	Like the plain butt, only continuous,long entire length of door. Cut to size with a hacksaw.	Extra-long or heavy doors.
Semi-Concealed	Like the plain butt except that one leaf is fastened to the back of the door, the other to the cabinet facing.	Partially overlapped cabinet doors with standard size lips or rabbets.
Concealed	Several types. Designed so that no part of the hinge is visible.	Cabinet doors that are flush to the cabinet facing. Involves mortising the door and the cabinet.
Offset	Like the plain butt only with one leaf offset for screwing into surface instead of edge.	Cabinet doors that are lipped, over lapping, or flush. Plywood edges.
Knife	Thin metal blades, designed so that only a thin strip of metal is visible.	Cabinet doors that are overlapping or flush. Mount to the top and bottom of the door.

SELECTING THE CORRECT SCREWS

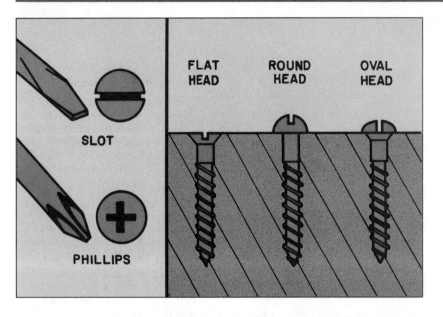

As shown here, there are several types of wood screws. They are identified by the grooves in the head (slot or Phillips) and the shape of the head (flat, round, or oval). The slot head screw is driven with a regular screwdriver, the Phillips head screw with a special screwdriver. Note how the flat, round and oval head screws meet the surface of the wood.

SPECIAL FRAMING FASTENERS

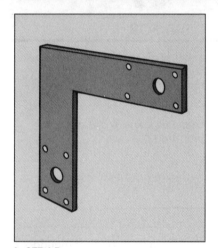

L-STRAP

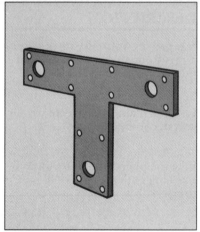

T-STRAP

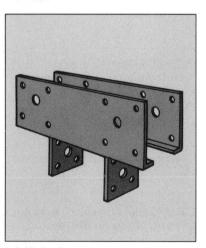

POST CAP

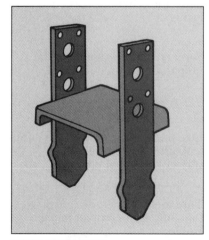

POST ANCHORS

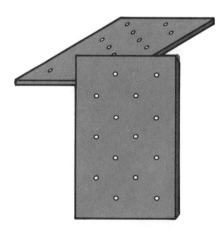

Nail-On Plates. Produced in a wide variety of shapes and sizes, these plates are routinely used to brace framing members and trusses. They usually have several holes for attachment with nails.

WOODWORKING MATERIALS AND HARDWARE

CHOOSING THE RIGHT BOLT

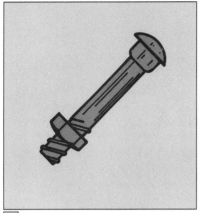

1 **Carriage Bolt.** Distinguished by its round head. Installed by tightening the nut which may be used with or without a washer.

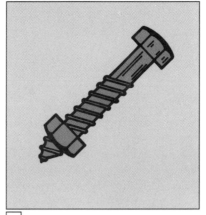

2 **Lag Bolt.** Hexagonal head; tightened by its head rather than a nut. Used for hanging cabinets or shelves at wood studs, often in conjunction with a brace or cleat.

3 **Machine Bolt.** Available with either a hexagonal or square head; tightened by its nut. Washers may be used at the head and the nut.

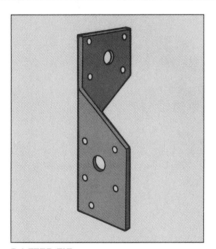

RAFTER TIE

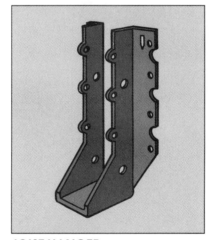

JOIST HANGER

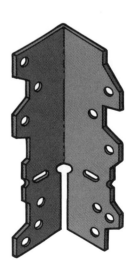

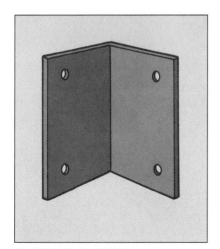

FRAMING ANCHOR

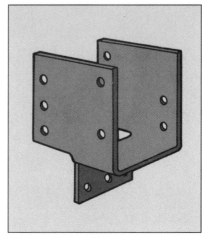

HEADER HANGER

Framing Connectors. Like nail-on plates, but made in special shapes for reinforcing posts, beams, and joists. Some have metal prongs to eliminate nailing.

4
Basic Woodworking Skills

Whatever type of work you're planning to do — an addition to your home, kitchen cabinets or a built-in bookcase — you'll need to know how to saw wood, make joints and assemble wood members. Some of these skills may look easy to the beginner, but they all take practice. How well you do them will be reflected in your finished work.

Both sawing and joinery can be done with hand tools. However, a table saw or radial arm saw and a jointer will do the job more quickly and accurately. **Whenever you use the table saw, radial arm saw, or jointer, always work with extra caution and follow the manufacturer's instructions. Power tools are dangerous if used improperly — the majority of woodworking accidents happen on these three tools.**

There are two basic saw cuts — *ripping* and *crosscutting.* All other cuts — miters, bevels, chamfers and compound cuts are variations of these.

Joinery is the 'cornerstone' of woodworking for in this process, wood members are fitted together in a variety of ways for strength, beauty and utility. Although there are many different types of joints, they are all variations of the five most basic types — butt, miter, rabbet, dado and groove. You should practice making these simple joints.

In the previous chapter, you were introduced to glues and fasteners. Here, you will learn how to apply glues, and how to properly install nails and screws. When you've mastered the techniques in this chapter, it's time to begin your own projects.

Start with a Mark . . .

All your projects will begin with precise measuring and marking, and the importance of this step cannot be emphasized enough. An error at this stage will not only waste time and valuable materials, but worse yet — result in a poor-looking project. All professionals 'measure twice, cut once' to avoid costly mistakes.

Marks can be made with a pencil or scratch awl. Whichever tool you choose, make sure that it is sharp and that you hold it as close as possible to your straightedge or square. This technique will give you the most accurate measurement. The marking process is shown below (left).

Continue with a Cut . . .

Whether you are cutting with a hand or power saw remember to allow for the saw kerf. The saw kerf is the amount of stock that is removed by the saw blade as it cuts through the wood. With this in mind, to make accurate cuts you should always cut on the 'waste' side of your awl- or pencil-mark. Study the drawing below (right); by always cutting in this way you'll be assured of an accurate cut.

Sawing Techniques. Depending on the position of the saw blade to the grain of the wood, there are two different methods of sawing: If the wood is perpendicular to the blade you are crosscutting (cutting across the grain) and if the wood is parallel to the blade you are ripping (cutting along the grain). Some tools are designed to do one method very well and the other only adequately. Such is the case with the radial arm saw which is a champion at crosscutting, and the table saw which excels at ripping. Hand saws (rip and crosscut) or a circular saw can do both crosscutting and ripping but with less precision.

Different saws obviously call for different techniques. If you're using a crosscut handsaw, first make a starting notch in the wood by pulling the saw blade

MARKING AND CUTTING WOOD

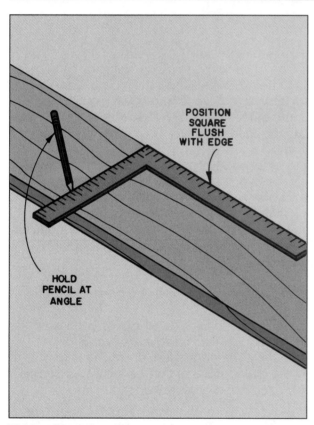

POSITION SQUARE FLUSH WITH EDGE

HOLD PENCIL AT ANGLE

Using a Square. Always use a straight, undamaged square to make your marks for cutting. Position one blade of the square flush to your workpiece and draw the cutline using the other blade as a guide. Hold your pencil or awl at an angle that creates a line close to the square's edge.

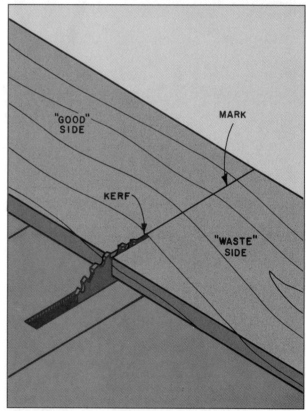

"GOOD" SIDE

MARK

KERF

"WASTE" SIDE

Cutting on the Waste Side. A saw cut should always be made on the *waste side* of your mark — not on the 'good' side, and not directly on the mark. If you plan to use the 'waste' material, always remeasure and mark it; that is, don't measure for many cuts in a row. By not allowing for the saw kerf, you'll make incorrectly sized wood members.

back across the spot where you want to make the cut. Next make several cuts using the part of the blade closest to the handle. Then, progress to long smooth back-and-forth strokes.

Sawing requires practice and patience to perfect, so you should spend time cutting scraps to acquaint yourself with the feel and operation of your hand saw, circular saw, table saw or radial arm saw.

Plywood and Other Sheet Materials. An important point to remember when cutting large stock is to always support the piece you're cutting. Get a helper and use saw horses to support the stock when you're using a handsaw or circular saw. If you're using a table saw or radial arm saw, get a helper and use roller stands and/or table extensions to help support the stock. If you don't support the stock properly, the saw blade will be "pinched" between the two pieces of stock. Also, the unsupported piece will break off and splinter as you near the end of the cut. A good way to keep a saw blade from binding is to use a kerf-keeper or wedge a thin strip of wood in the beginning of the cut to keep it open. Most table saws and radial arm saws are equipped with mechanisms that prevent the saw blade from binding. **If a piece of stock binds on a table saw or radial arm saw a dangerous kickback can occur.**

Avoiding 'Tear-Out'. Wherever saw teeth exit the stock, there is a chance of creating splinters or 'tear-out'. If the saw blade you're using tends to create splinters, make sure that they will appear only on the side of the stock that won't be visible on the finished project.

Other ways to avoid tear-out are to apply masking tape where your cutline is to be or to score a line on the back side of the stock with a knife. Yet another method is to back up the stock with a piece of scrap wood and cut through both pieces at the same time.

Using the Correct Saw Blade. You can avoid problems just by using the proper blade for the material you're cutting on your table saw, radial arm saw or circular saw. Though a wide variety of blades exist, the following are the most common:

A *hollow-ground blade* is the best type to use for crosscutting, while a *rip blade* is used exclusively for ripping. Use a *plywood blade* to keep the laminated layers of plywood from tearing out. A carbide-tipped *combination blade* is a good investment because it can be used for cutting most materials, plus it will stay sharper for a much longer period of time.

CAUTION
Always wear proper eye protection when using power tools (pages 3 and 4). Also, be sure to keep your hands out of the danger zone (page 5).

Simple Cuts. Four simple cuts that can be made with the radial arm saw and/or the table saw are the *crosscut, rip cut, miter cut,* and *bevel cut.* The crosscut and rip cut are the easiest to make although they do require precision. Miter and bevel cuts are made by adjusting the saw's settings.

Joining — Connecting and Securing Wood Members

Whether you're building a cabinet or a wall, you will need joinery — the connecting process which is vital to all projects. Although there are many different types of joints, learn the simple ones first. All the rest are merely adaptations of these.

When choosing which type of joint to use, you should make your choice based on several factors: the strength of the joint, its appearance, the tools you'll be using, and the degree of skills you possess. The chart on page 44 will help you to make your selection. Follow the steps on pages 46 and 47 for making a dado or a groove with a router, and those on page 45 for making a rabbet with a table saw.

Once you have mastered simple joints, you may want to advance to combination joints — mortise and tenon, lap, tongue and groove, and lock. Shown on page 48, these are generally stronger and more attractive. As a caution to the beginner, however — simpler is better. Make sure that you have become accomplished at making simple joints before tackling the more difficult types.

Pay Attention to Grain. One thing that people often forget is that wood has living qualities; that is, it 'breathes' by expanding and contracting in accordance with the temperature and humidity. These changes are not noticeable along the length of the grain in a board, but that same board will move up to ¼ inch for every 12 inches of width across the grain. The movement is insignificant where small boards (less than 3 inches wide) are joined together, but larger pieces will swell and 'pop' the joint, or shrink and split out.

Some joints are especially designed to accommodate the movement of wood. The entire purpose of frame-and-panel joinery, the kind regularly used for cabinet doors, is to give the panel room to breathe without distorting the frame. The panel sits loose in slots in the frame, free to swell or shrink.

Wherever you can, align the wood grains of adjoining boards so that they expand and contract in the same directions. This is especially true if you live in an environment where temperature and humidity changes are radical.

Reinforcing Joints. Normally, a well-made joint does not have to be reinforced. But for projects that will have medium to heavy loads, and especially for workpieces that are joined with butt joints, reinforcing is a good idea if not a necessity. The main purpose of a reinforcement is to increase the gluing surface of a joint.

Two of the most popular types of reinforcements, *dowels* and *splines* are 'invisible', that is, they are hidden inside the joint. A third type, the *glue block*, is more noticeable but it's fine for 'out of the way' places such as the insides of drawers. These three types of reinforcements are shown on page 49; the process of using dowels for reinforcement is also shown on page 49.

Another way to ensure a joint's strength is to reinforce it with nails or screws. Both of these are like mini-clamps in their function, so they are always installed after gluing.

Assembly — Uniting and Securing Wood Members

The assembly process is critical to the success of your project. You should 'dry fit' wood members together and 'feel' how they fit to each other — joints should be neither too loose nor too tight. Also, check them to see if they fit true. If there is the slightest irregularity, you must correct it.

Use chisels to smooth and adjust rabbets and dadoes, and use sandpaper on other kinds of joints. A little bit of work at this point will pay off greatly in the long run. Be sure to constantly check the joints when working on them — you don't want to make them worse by removing too much stock. Shown on page 51 are two smoothing techniques for fitting a tenon to a mortise.

Bonding — Good Gluing and Clamping. As mentioned previously, glues differ in their uses, so you should always pay attention to the manufacturer's recommendations. Check to make sure that you are using the proper glue for your project and that you have adequate ventilation.

MAKING BASIC SAW CUTS

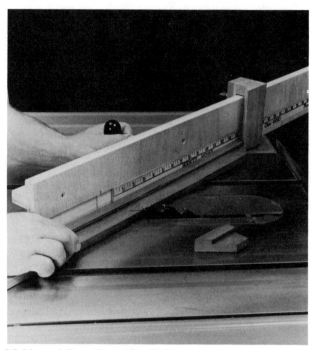

Ripping with a Table Saw. Make a rip cut by first placing the stock on the front edge of the table, good side facing up and the edge snugly against the rip fence. With the saw blade guard down, use your left hand to hold the stock against the fence. With your right hand, push the work slowly forward until your hand approaches the fence; at that point, hook your fingers over the fence. Continue holding the work against the fence with your left hand and feeding it with your right. Note that the cut is made parallel to the wood grain. Another way to make a rip cut with the table saw is to use a featherboard to hold the stock against the fence, and a push stick or push block to advance the stock.

Making a Miter Cut with a Table Saw. Mark the cutline on the stock. Set the miter gauge to the correct angle. Place the stock snugly against the miter gauge and position the stock so the blade falls on the waste side of the cutline. Switch on the saw and cut *very slowly* while holding the stock snugly against the miter gauge. If necessary, use a miter gauge extension. A miter cut also can be made with the radial arm saw.

CAUTION If you're making a narrow cut which requires that the rip fence and saw blade be close together, and your hands cannot be safely positioned, use a push stick to feed the stock.

When applying glue, be sure to coat the surfaces correctly. Some glues work best when applied to only one surface; others should be applied to both surfaces. Some should be applied in a thin coat, others heavy. Even with these differences, there are a few rules that need to be followed when using all glues:

First, always coat the *entire* surface or surfaces; don't simply 'bead' the glue on and expect it to spread evenly when you clamp your work. A common method is to apply glue in strips and then rub surfaces together in a circular motion to distribute it evenly. Use a paint brush to spread glue on large and hard to reach areas.

Second, remember that glue of any type will soak into the end of a board at a much greater rate than it will into the edge or surface. In general, apply more glue to the ends to allow for this 'loss' of glue. Unless the glue you are using will bond very quickly, allow a little time for it to soak into the wood before clamping.

Clamping is a misunderstood skill. Some people think that a clamp should press joints together to the point of extreme pressure. This is not true because in doing so you'll create a 'starved' joint — a joint that has insufficient glue for providing holding power. Excess pressure might also mar the wood. Use just enough pressure to keep the workpieces from moving. Always wipe away excess glue immediately with a wet rag. Otherwise, it's almost impossible to remove glue once it dries.

Making a Bevel Cut with a Table Saw. Use the rip fence as a guide for making a *rip bevel* cut. Depending on what type of table saw you own, tilt the saw blade or the table to the correct angle. Place the edge of the stock snugly against the rip fence. Align the stock so the blade falls on the waste side of the cutline. Switch on the saw and cut very slowly while holding the stock snugly against the rip fence. Use the miter gauge to hold the stock when making a *cross bevel* cut. Tilt the saw blade or the table, and set the miter gauge at the correct angle.

Crosscutting with a Radial Arm Saw. To crosscut with this tool, first place the work against the fence. Turn on the power, then pull the saw blade forward, through the material, to make the cut. Note that the cut is made perpendicular to the wood grain.

CAUTION Always wear proper eye protection when using power tools (pages 3 and 4). Also, be sure to keep your hands out of the danger zone (page 5).

CHOOSING THE CORRECT JOINT

TYPE	DESCRIPTION	USES AND BENEFITS
Butt 	Two or more members joined end to end, end to edge, face to edge, or edge to edge.	The weakest but acceptable when members are joined with glue and reinforced. Used for face frame construction.
Miter 	Two members cut at an angle across the wood grain and joined.	Attractive but not very strong; needs reinforcing. Used for room and door moldings, cabinet door frames, and facings.
Rabbet 	A wide notch cut in the edge or end of a board in which another member sits.	Strong. Often used in cabinet construction — for cabinet backs and drawers.
Dado 	A wide cut in the surface of the stock, across the grain.	Strong; withstands stress from several directions. Often used for drawers and for supporting shelves in cabinets and shelving units.
Groove 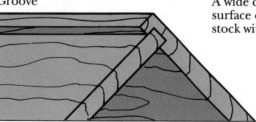	A wide cut in the surface or edge of the stock with the grain.	Strong; withstands stress from several directions. Often used in cabinet construction.
End Lap 	Two members with wide rabbets cut in the surface.	Strong; used in cabinet construction for facings and door frames.

CAUTION

Always wear proper eye protection when using power tools. Also, be sure to keep your hands out of the danger zone (page 5).

1 Making a rabbet joint requires two passes, one on the surface of the stock and another on the edge. Make the surface cut first. Remove the upper saw guard, adjust the table height, and position the rip fence the desired distance away from the blade. Make the first pass. Keep your hands out of the danger zone — use a push block to move the stock past the blade.

CAUTION

In order to make a rabbet with a table saw, you need to remove the saw blade guard. Work with extra caution whenever you work without the guard.

2 For the second pass, turn the stock on its edge so that the waste will be on the opposite side of the blade from the rip fence. In this way, you'll avoid getting the waste stock caught between the rip fence and the blade — a dangerous situation because the waste could kick back at you. Keep your hands out of the danger zone, and use a push block to move the stock past the blade. If the stock is wide, mount a high fence extension to the rip fence to help support the stock safely.

1 Adjust the router for the depth of the channel. With the power off and the router unplugged, insert the bit into the collet and then tighten the collet nut. Sit the router upright on the stock to be cut. After loosening the locking screw, turn the depth-adjustment collar until the tip of the bit makes contact with the stock. Set the depth scale to zero and then move the router to the edge of the stock. Turn the collar to the desired depth; tighten the locking screw.

CAUTION

Always wear proper eye protection when using power tools. Also, be sure to keep your hands out of the danger zone (page 5).

3 Position a router guide — straight-edged piece of wood — along the line you have just drawn and clamp it in place at both ends with C-clamps or handscrew clamps. Check to make sure that the guide touches your line at all points and is, indeed, straight.

2 Measure and mark for the perimeter of the channel and then place the router on the stock so that the bit lies flush with one edge. Make a mark where the outside edge of the router base meets the stock on one end. On the other end of the stock, make another similar mark. Connect the two marks with a straightedge and awl. The line you have made is for the router guide.

4 Align the router next to the guide so the bit nearly touches the edge of the stock. Turn on the power and gradually ease the router forward (away from you) so the bit makes contact with the stock. Push the router at an even speed; moving it too slowly will burn the stock, too quickly will drag on the motor. When you reach the end of the channel, carefully lift the router clear of the stock and turn off the power. To make a wider channel, reposition the guide and repeat the process.

To make a groove in the edge of a piece of the stock, follow the same procedure.

COMBINATION JOINTS

Combination joints are merely simple joints arranged in combination with each other. In order to make these joints, you should first master the simple joints — from which they are constructed.

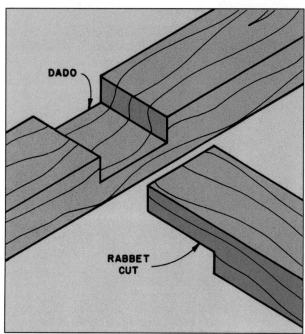

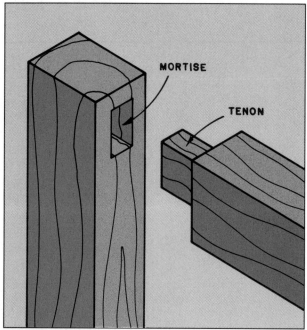

Mid-lap Joint. This strong joint is excellent for cabinet construction. Consisting of a rabbet and dado, it requires skill and precision to make.

Mortise and Tenon. One of the oldest and strongest joints; often used for leg attachment and reinforcement. The mortise is a dado that's closed on both ends. The tenon is four rabbets on the end of a board.

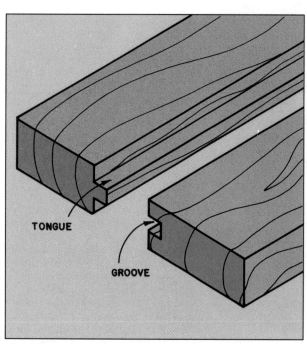

Tongue and Groove. A good joint for edge-joining stock for wainscotting and house siding. Composed of a narrow tenon and a matching groove.

Lock Joint. Extremely strong and attractive but very difficult to make. A combination of dado, groove, and tenon joints.

REINFORCEMENTS FOR JOINTS

Reinforcements increase the gluing surface of a joint.
Shown here are the three most common types:

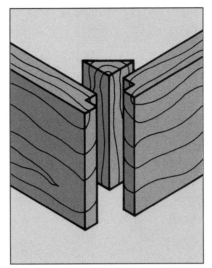

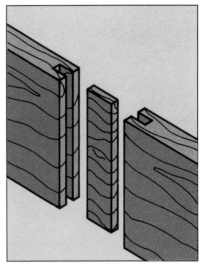

Dowels. Descendants of the wooden pegs used before nails, dowels are round pins with extremely good holding power. Typically made with flutes (grooves) or spirals for acceptance of glue, they must be fitted with care. Matching holes are drilled.

Glue Blocks. Used in many situations but especially useful where there is minimal gluing surface. Normally used to reinforce concealed joints.

Splines. Splines provide strength to edge and corner joints. They are thin flat sections of hardwood inserted in narrow grooves. The grooves can be cut with a table saw.

DRILLING FOR DOWELS

1 Begin by making your measurements for dowel holes. Clamp the boards to be joined, surface-to-surface, as shown. Using a square, mark the hole locations across the edges.

2 Use a doweling jig, a power drill accessory, to create the holes. Set the jig for the correct size hole and carefully align it over your mark. Securely clamp the jig to the stock and drill the dowel hole.

BASIC WOODWORKING SKILLS

1 Although methods vary slightly with the type of glue used, most glues should be applied to both surfaces, as shown here. Glue can be applied with a brush, tube injector or other available accessories. Check the manufacturer's instructions for whether to apply glue in a thick or thin coat. A quick way to spread glue is to rub the surfaces together in a circular motion.

2 Clamp joined wood members together only until the glue *begins* to ooze. If you clamp too tightly, you'll cause a 'starved' joint — a joint with too little glue.

3 Quickly wipe away excess glue with a wet rag. Since this will raise the grain slightly, you must remember to sand this area well when preparing for the finish.

BASIC WOODWORKING SKILLS

Fasteners — Nails and Screws. The process of driving a nail compresses wood fibers and makes them 'cling' to the fastener. Likewise, a screw is constructed so that its thread pulls the wood fibers in the direction of the screw head thereby clamping the wood.

Nailing must be done carefully; you should always be aware of the wood grain pattern and where nails have been previously driven. Nails with sharp points hold better than blunt ones but a disadvantage is that they tend to split the wood. To avoid splitting, you can blunt a nail by striking its tip with your hammer, or you can drill a pilot hole into the wood before driving the nail. All of these nailing techniques, and a few more, are shown on pages 52 and 53.

Using screws is a little more complicated but the end result is a stronger joint. Remember also that screws may be installed *without* glue so that you can disassemble your work. See pages 33 and 36 for which type and size of screws to use.

FITTING A TENON TO A MORTISE

1 Use a rasp to round the edges of a tenon to fit a mortise. Make several passes on either side holding the rasp in a diagonal direction to the tenon

2 Chisel away the sides of the tenon by hand until it fits the mortise properly. It should be snug enough so that the parts can be fitted together firmly but without the use of a mallet.

Blunting the Nail End. To avoid splitting the stock when nailing, lightly tap the pointed nail end with a hammer to blunt it.

Creating a Pilot Hole. A pilot hole is useful when working with stock that tends to split. If you can't find a drill bit of the correct size, use a nail in the chuck of your drill to create a pilot hole. Use a size smaller than the one you intend to drive.

Using Nail Sets. Instead of crushing the stock with the last few hammer blows, maintain a smooth surface by setting finish nail heads about 1/16 inch below the surface.

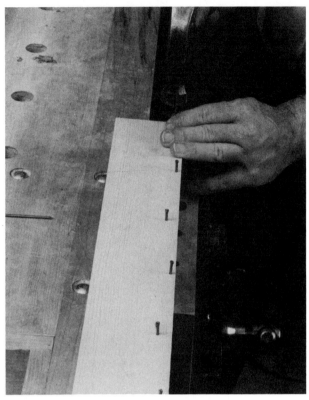

Placing Nails. Try to avoid placing nails along the same grain lines since this can cause splitting. Instead, stagger them as shown.

Removing Nails. To remove a nail, wedge the claw around the nail's shank and rock the hammer backward. To protect the stock's surface, place a piece of scrap wood under the head of the hammer.

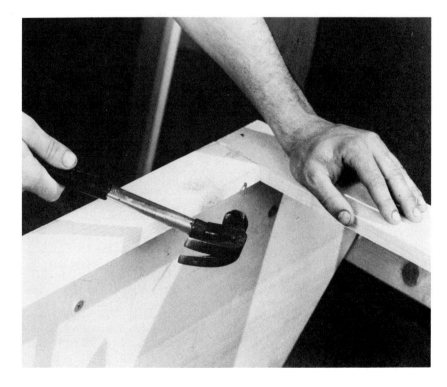

Toenailing. To make corner joints snug, drill diagonal holes and then toenail members together as shown.

Generally, screws should not be driven until you have first drilled pilot holes; these are made using a drill with a small diameter bit. Exceptions to this rule are woods that are very soft and screws that are small and short. In these instances, pilot holes may be made simply with an awl or other sharp instrument. The best way to determine if you need a pilot hole is to try driving a screw into the wood without one. If you meet resistance, then a pilot hole is needed.

Pilot holes are shaped differently depending on whether the screw head is to be raised above the wood surface, level to it, or sunk below it. The terms for these are called, respectively: *simple pilot hole*, *countersunk pilot hole*, and *counterbored pilot hole*. They are shown below.

All three of these holes need to have at least two different diameters. One, the lower part of the hole, is for the portion of the screw that is threaded; its diameter should be the same diameter as the core of the screw minus the threads. The other, the upper part of the hole, is slightly larger to accommodate the unthreaded portion of the shank.

Countersunk screws, of course, require an additional V-shaped hole at the top to accommodate the screw head. Counterbored screws go one step further, requiring that the V-shaped hole be recessed. A special drill bit like the one on the right is recommended for drilling the pilot hole, countersinking and counterboring in one easy step.

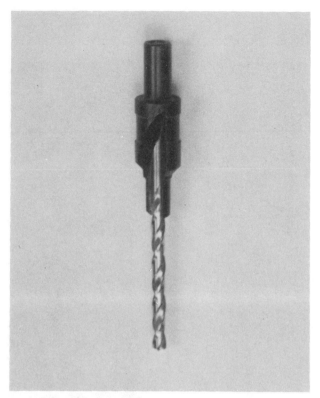

An aid to drilling the pilot hole, countersinking and counterboring operations is this special drill bit that will perform all three functions.

DRILLING PILOT HOLES FOR SCREWS

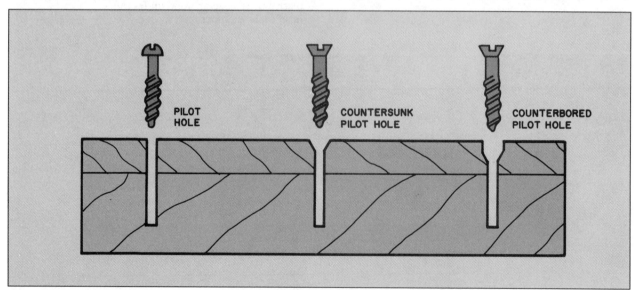

PILOT HOLE

COUNTERSUNK PILOT HOLE

COUNTERBORED PILOT HOLE

Shown here are three kinds of pilot holes: a simple pilot hole, a hole for countersinking and one for counterboring. The head of the roundhead screw is visible above the surface of a simple pilot hole. The top of the countersunk flathead screw just meets the surface. The counterbored screw is recessed below the surface — for insertion of wood putty or a wood plug.

To make a simple *pilot hole*, begin by drilling a lead hole — the same diameter as the screw's core minus its threads and to a depth slightly shorter than the length of the screw. Next, make a body hole — the same diameter as the screw's shank and slightly shorter than its depth. To make a countersunk pilot hole, follow the above instructions and then use a special cone-shaped bit for the needed cavity. To make a *counterbored pilot hole*, do all of the above; then use a counterboring bit to drill to the correct depth for the deep hole at the top.

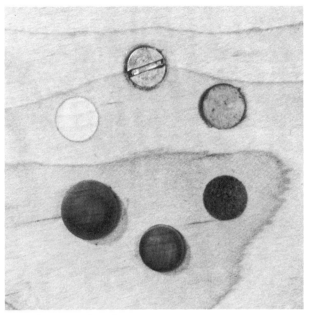

Disguise screw heads in a number of ways. Clockwise from the top are: *Countersunk* (but uncovered) screw, *wood dough, sawdust mixed with glue, dowel, button,* and *plug.* The plug, considered to be the most attractive, is made out of the same material as your stock. A special plug cutter is used.

Once you have drilled your pilot hole, it's time to drive the screw. In order to do this properly, use the correct size and type of screwdriver — blade or Phillips. One that's too large or too small for the screw slot(s) will either burr the screw head or slip off and gouge your work.

If you meet resistance when installing a screw, remove it and enlarge the hole. Screws are much easier to drive if their threads have been rubbed with soap or wax. Most importantly, keep the screwdriver straight and squarely seated in the screw slot.

Hiding the Evidence — Covering the Heads. Screws are excellent devices for holding wood members together, but depending on your project, you might not want them to show. There are special ways of covering screw heads. These range from filling the holes with wood putty to gluing various pieces of wood into place, such as dowels, buttons, and plugs. Glue dowels in the holes and sand off the excess. The only drawback to this is the dowel may not match the wood. Therefore, plugs and buttons are preferred since they can be purchased or made from the same type wood as your projects.

Home
Carpentry

5

Basic Residential Construction

When the time comes to build an add-on to your home, you will need to understand both how your present home is constructed and enough about construction to be able to design the add-on. Obviously you won't become a master overnight and you won't be able to work with difficult designs. You will, however, learn the basic terminology of the trade, so you'll be able to plan the project and talk intelligently to subcontractors and suppliers. Most of all you'll gain the necessary confidence to tackle simple structures and tasks yourself. That's what this chapter is all about.

First, you'll see the anatomy, the skeleton, of a structure beginning with the foundation. The foundation is extremely important; without a good solid foundation a house is subject to settling or movement, both of which will affect the framing members of the structure. The elements of foundations are detailed.

Next, you'll see how the outside wall framing determines the basic design of the structure. Two popular types are shown here: platform framing and balloon framing. You'll learn how to distinguish between the two — an important consideration when building an add-on. If you're building a stucco or brick add-on, you'll have to study additional research material — available at the library or your local home improvement center. Basic information about roofs is also presented so that you'll know how to calculate the slope of a new roof.

Finally, you'll walk through the basic steps of planning for your add-on. Drawings, permits, and contractors are all discussed. You'll learn what to do first and what to do on your own. All of this information is *your* foundation for the next chapter — the hands-on depiction of how it's all actually accomplished.

Structure Anatomy — An Overview

A look around your neighborhood will probably reveal houses with a variety of add-ons — room additions, garages, carports, patio enclosures, porches and sheds. Each of these is a unique structure. But don't let this uniqueness frighten you. All structures contain the same basic elements — footers, foundation walls, concrete slabs or wooden subflooring, wall framing, and roof framing. Any given structure can be a mixture of different variations of these elements. If you learn only the basic elements, you'll be able to figure out how various add-ons are constructed.

Begin at the Bottom...

Structures are placed in the ground for equal distribution of their weight, and to prevent lateral movement or racking. In a typical structure, the parts that take care of these functions are the footers and foundation walls. The footers act like feet and the foundation walls act like legs to support the structure.

For most add-ons, concrete slabs or wooden subflooring systems are the most practical, economical and most often used types of foundations. This text will describe how both types are constructed. Slabs are not recommended in certain geographical areas

ELEMENTS OF FOUNDATIONS

The first and most common elements of foundations are the *footers and foundation walls* which are located at the perimeter of the structure. Footers are wide concrete bases

upon which the foundation walls are built. There are two basic types of foundations — slabs and subflooring systems. Both require footers and foundation walls.

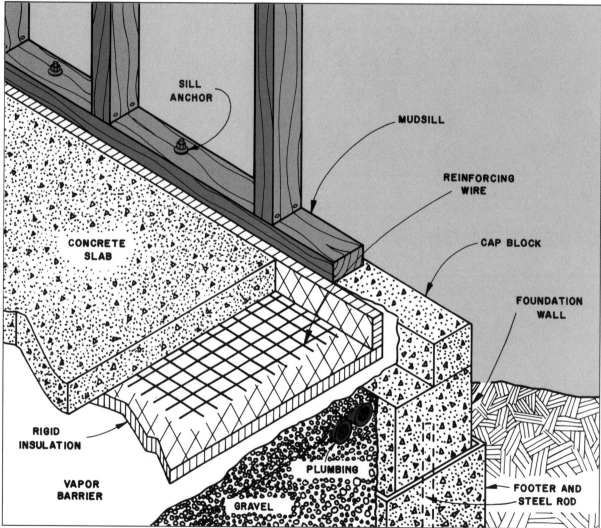

Used in conjunction with footers and foundation walls, the slab is a flat concrete pad. *Gravel, plumbing, a vapor barrier, rigid insulation, reinforcing wire,* and *concrete* are essential components.

where the frostline renders them unsuitable. In these areas, check the local building codes for the proper type of foundation to use.

The Foundation — Footers and Walls. A sound foundation is essential. In order to create one, you or the subcontractor must first make sure that the terrain is suitable. If soil conditions are poor, or if the foundation is not properly designed, then the structure will be subject to uneven settling.

The procedures for constructing the footers and foundation walls are shown on pages 62-64. Forms are necessary for pouring the concrete for the footers. These forms must be built and precisely positioned.

Most footers require steel reinforcing rods. Follow local building codes.

Pouring the concrete is no easy process; temperature and timing are crucial factors for doing it correctly. You may want to let a subcontractor lay out the footers and pour the concrete. Then you could erect the concrete blocks for the foundation walls yourself. It's recommended that you have some experience doing masonry. Laying concrete block must be done very carefully with a stringline guiding you all the way and constant checks of level. The corners are set in first, then the rest of the blocks are filled in.

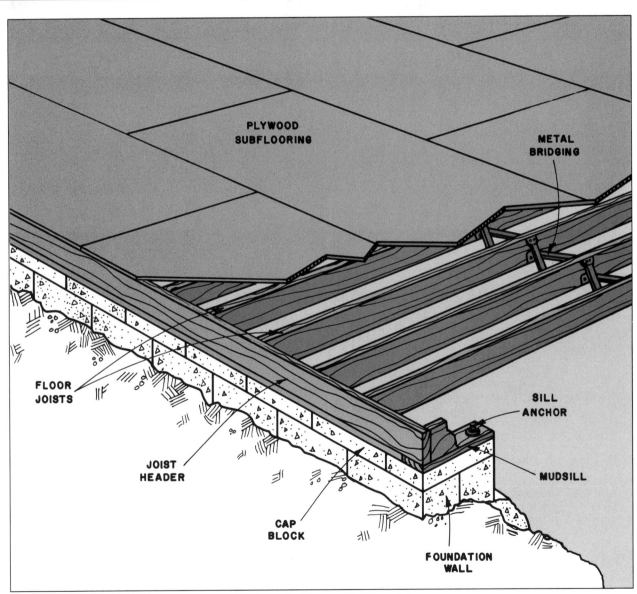

Elements of the subflooring system are the *mudsills* which attach with *sill anchors* directly atop the *cap blocks.* Resting on these are the *joist headers* and the sturdy *floor joists* which are often braced with metal *bridging. Subflooring* — sheets of plywood — is laid over the joists.

BUILDING A SOLID FOUNDATION

1 Before you begin any excavation for a home addition, be sure to notify your utility companies. Here, the power company 'painted the ground' to designate where underground wiring exists.

2 Pour the footers. As soon as they set, mark the footers with chalk-lines to define the perimeter of the concrete block walls.

5 Measure for and drill the holes in the cap blocks for the sill anchors. Use a carbide-tipped masonry bit in the drill.

6 Partially fill the cavities in the blocks with coarse gravel. Fit the sill anchors in the cavities, holding them upright. Add cement to the cavity. The locations of the sill anchors are determined by code. These are the anchors that will secure the mudsill to the foundation wall.

3 Determine the corners and run a stringline to set them up. After the corner blocks are set, the string may be run from them.

4 Next, fill in the center sections with blocks.

7 Fit the drilled cap block over the sill anchor.

8 Make careful checks for level at the corners and periodically during construction of the wall. Even up the blocks by tapping them lightly with a hammer.

BASIC RESIDENTIAL CONSTRUCTION

9 Use a special tool, a *convex jointer*, to make slight grooves in the mortar between the blocks.

10 Two-inch styrofoam is set in next to the interior side of the wall. Here the chalkline is being snapped for the position of the mudsill.

11 This particular room addition calls for the removal of the exterior wall, so the builder is demolishing the wall and using the brick for filler before the gravel is added.

12 Add a layer of gravel inside the foundation. Install the plumbing at this point. Follow this with the vapor barrier and the rigid insulation. On top of the insulation, lay reinforcing wire. Then pour the concrete.

The foundation walls should extend no less than 8 inches above the finished grade. This assures good drainage and moisture protection plus it aids in termite control; pest tunnels can be observed easily if they are present.

The top row of blocks, called *cap blocks,* are solid. These act as termite barriers unlike the lower ones which have holes in them. Before the cap blocks are set, they are drilled at strategic points for the insertion of the *sill anchors.* Their position is determined by local building codes.

An alternative method to cap blocks is filling the top row of blocks with concrete. While the concrete is still workable, the sill anchors can be seated directly in the concrete — per your local building codes.

Building a Slab.   The slab is constructed of layers of various materials laid within the foundation walls as shown on page 60. After the foundation walls are built, a layer of gravel is added. If your add-on requires plumbing, have a plumber set in the rough plumbing at this time. You must also put down a vapor barrier. This is crucial as a guard against moisture and temperature. The vapor barrier — usually 6-mil polyethylene or a similar material — is positioned on top of the gravel. In some areas, you should add a layer of rigid insulation. Insulation

PLATFORM FRAMING

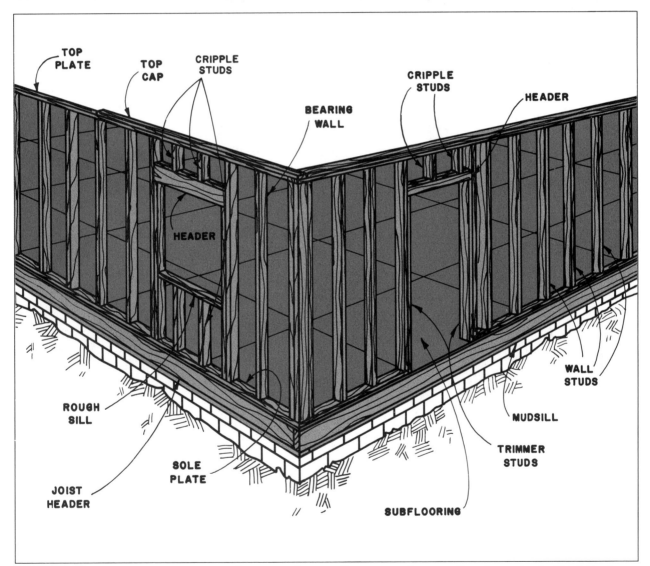

This popular method of framing requires that the builder erect wall sections above the subflooring *one story at a time.* Typically wall sections are assembled on the floor, then raised into position. After second-floor joists and subflooring are in place, second-story walls are erected in a similar manner. Components are *studs* and *plates.* The *sole plate* provides the base to which studs are attached. The *top plate* receives the top of the studs and the *top cap* rests on it. (Together, these are sometimes referred to as the *double top plate.*) Small studs above openings are called *cripple studs;* these rise above the horizontal *headers.* The members around openings are called *trimmer studs.* A window base is a *rough sill.*

BASIC RESIDENTIAL CONSTRUCTION

prevents excessive heat buildup in warm climates and reduces heat loss to the outside in cold climates. On top of the vapor barrier, lay metal 'road mesh' for reinforcement. Finally, pour the concrete.

Building the Subflooring. The construction of the subflooring is shown on page 61. The mudsill is the first wood member resting on top of the foundation. The mudsill is attached to the sill anchors that were imbedded in the cap blocks. The mudsill is the starting point for the floor framing as well as the wall framing. Other subflooring members include floor joists and joist headers.

When platform framing (page 65) is used, a joist header is employed on the perimeter of the mudsill.

To it are attached the floor joists that span the structure. Floor joists may be long and continuous, or they may be overlapped where they meet a central supporting girder. Floor joists should be reinforced with metal or wooden bridging. Once all the framing is completed, then the plumbing, wiring and insulation are installed. Finally, the subflooring is laid, usually in the form of sheeting material such as plywood.

When the method used to build the structure is balloon framing (below), floor joists are nailed to the mudsill in the same manner as for platform framing except that no joist header is employed. Between the floor joists and around the perimeter of the structure, short fire blocks are installed to stop airflow in

BALLOON FRAMING

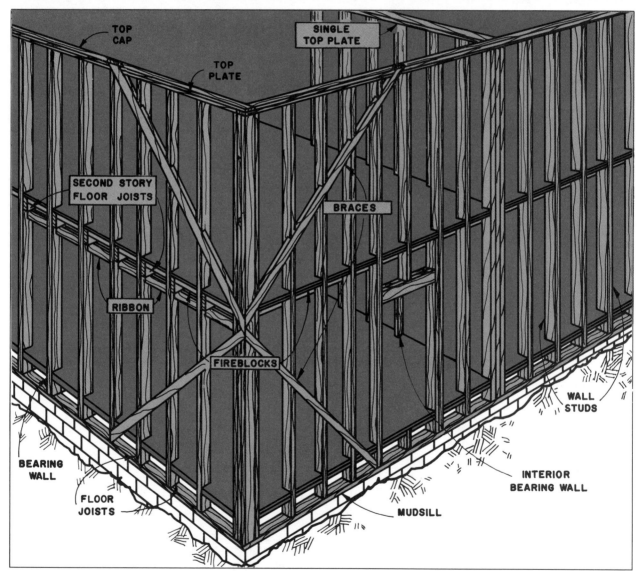

This type of framing consists of tall bearing wall studs that extend *in one piece* from the mudsill to the double top plate at the roof. First-floor joists are nailed to the mudsill. Second-floor joists span from the supporting *ribbons* to the *single top plate* which rests atop the interior bearing wall's 'short' studs. Diagonal *braces* add stability while *fire stops* help to prevent the spread of fire plus provide a base for nailing the subflooring. Balloon framing is best accomplished by a skilled crew geared to work at heights.

BASIC RESIDENTIAL CONSTRUCTION

case of fire. Metal bridging should also be used to reinforce the floor joist. Sheets of plywood are nailed to the floor joists and to the fire blocks. The floor framing of the second story is dependent on short ribbons for support of the floor joists.

Slab houses have a mudsill, but no floor joists. The slab serves as the subflooring and the wall framing starts at the mudsill.

Pest Prevention. No matter what type of foundation and subflooring combination you choose, you should be aware of the problem with termites. These insects can do considerable damage to the wood in your house — and they're difficult to detect. There are several ways to discourage them.

If your area is highly infested with termites, then you should use all chemically-treated lumber for your project. In less active areas, only the first-floor sills, joists and subfloors need to be treated. Another method calls for termite shields, metal strips that are fastened between the foundation wall and mudsill. These force the termites to build shelter tunnels so they can be easily spotted and destroyed. Other deterrents are to provide good drainage away from your foundation and avoid building on ground that is backfilled with wood scraps.

If your area permits, have the foundation area 'poisoned'. This chemical treatment will provide many years of protection.

Constructing the Shell — Two Methods

The two most common methods of framing the exterior walls of a house or any structure are *platform framing* and *balloon framing*. The platform method involves 'stacking the layers' of the building while the balloon method involves framing the entire outside first and then adding the floor structures. Although one-story houses can be either platform-framed or balloon-framed, the differences between the methods become more apparent when examined in two-story structures. Generally platform framing is the most common but the exceptions are homes built before 1930 and two-story stucco, brick, or masonry houses.

Platform framing, as shown on page 65, is certainly the preferred method for the first-time builder. Materials are easier to handle and sure footing is provided at the second story by the subflooring.

Balloon framing (page 66) also offers some advantages. Because the vertical studs are continuous, a very stable exterior surface is formed for masonry sidings and wall coverings. Shrinkage and settling is more uniform because fewer framing members are used. Difficulties with balloon framing include the hazard of working at heights and the problem of finding quality studs that measure 18 to 20 feet.

Identifying the Framing. If you're going to add onto your present structure, you'll need to know which method of framing was used to construct it. This is very easy to do. Go into your basement or crawlspace and locate the mudsill. If you find floor joists and wall studs paired together, your house is constructed with balloon framing. If the joists alone are set on the mudsill, the house is platform-framed.

Topping It Off — The Roof

Roof styles can be very elaborate and complicated but for the beginner a basic knowledge of a few standard types will be adequate. Three of these are shown below; choose among them depending on your

THREE ROOF STYLES

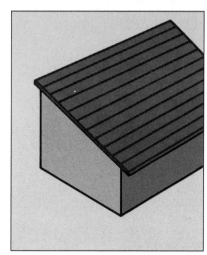

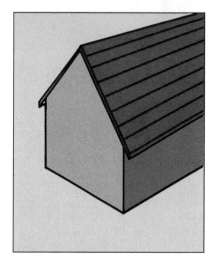

Flat. The simplest type. The ceiling joists support the covering materials. Usually made with a slight slope.

Shed. Similar to the flat roof except it is pitched in one plane. Often called a *lean-to*.

Gable. Consists of two straight slopes that rise and meet at a ridge. Named for the triangular wall or gable which is formed at the end.

project. For example, a flat roof will suffice for a storage barn; a shed roof will be appropriate for certain add-ons, and a gable roof will make an attractive garage covering.

Sloping the Gable Roof

If you live in a sunny dry climate you'll probably be interested in a gently sloping roofline. If you're building in an area where snowstorms are common, you'll want to have a more radical slope to your roof. Guidelines for roof slope will be available at your local building department.

Planning for the construction of a gable roof involves understanding the roofer's terminology and some general design principles. Study the illustration on page 69 to familiarize yourself with gable roof framing components.

To determine the length of the roof slope a knowledge of basic geometry is helpful. Suffice to say that the run and rise form two sides of a right triangle. Once these two measurements are determined, you can calculate the total distance of the slope. Units are the components that make for an easy conversion to carpenter's language. The unit for expressing the run of the roof is always 12 inches. Thus, when a pitch is described as '6 in 12', it means that for every foot of run, the roof rises 6 inches.

The only tool you'll use to step off a common rafter is a regular framing square. The framing square is placed along a common rafter with the body of the square always at 12 inches and the tongue of the square at 6 inches (or whatever figure is desired). Stepping off a rafter will not only determine the length of the rafter from the top plate to the peak but it will also give you the pattern for making the plumb cut, tail cut and bird's mouth. Follow the instructions on page 71 to step off a rafter.

ROOFER'S TERMINOLOGY

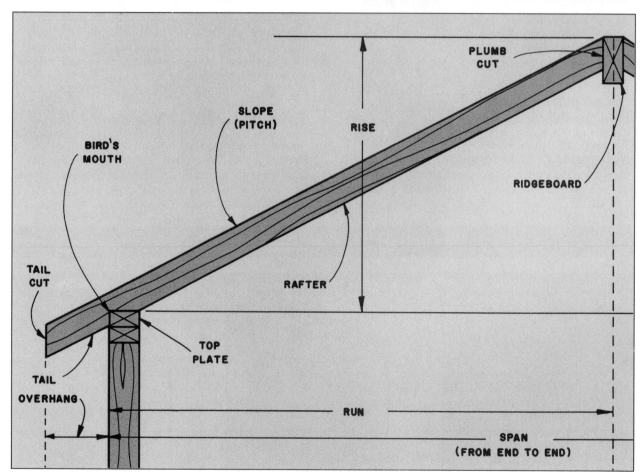

Span. The distance between the outside edges of the top plates.
Run. Half the span, from the point directly below the roof peak (the *midspan*) to the outside edge of the top plate.

Rise. The vertical height from the top plate to the peak, measured at the midspan.
Slope. The *pitch* or angle at which the roof is slanted from peak to tail.
Tail. The portion of the rafter that extends beyond the building.

Tail Cut. The cut in the rafter at the end of the overhang.
Plumb Cut. The cut in the rafter where it meets the ridgeboard.
Bird's Mouth. The notched cut in the lower end of a rafter where it meets the top plate.

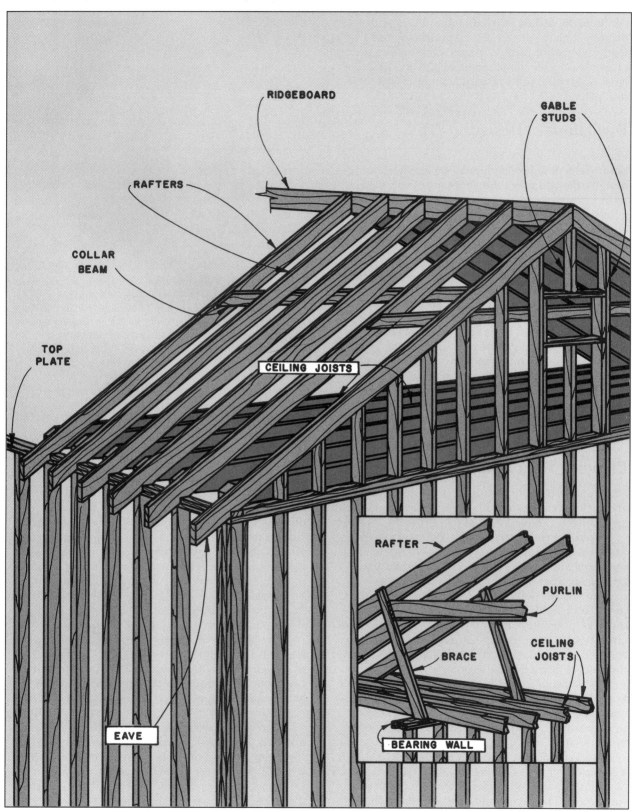

Matching pairs of evenly spaced *rafters* rise to meet at the central *ridgeboard*. Where the rafters meet the double top plate, they are notched to fit and extend over the side. This overhang is called the roof *eave*. The ends of ceiling joists are trimmed to match the slope of the rafters. When rafters span a long distance they should be bridged with *collar beams* or supported with horizontal *purlins* that are attached to vertical *braces*. The *gable studs,* attached to the top plate, are angled to meet the rafters.

A Roofing Alternative. One way to speed up the job of measuring and cutting roofing components is to use *trusses*. These triangular structures combine the ceiling joist, common rafters, and diagonal bracing all into a single unit. Normally made at a mill or factory, trusses, even smaller ones, take at least three people to handle them safely. For large ones you'll need heavy machinery to lift them into place.

Deviations in Design

Platform framing and balloon framing are the norm but if you feel somewhat limited by standard house designs, then consider the alternative of post and beam construction.

Post and beam framing is a throwback to the way barns were built. Instead of the light 'two-by' lumber placed close together in traditional framing, post and beam incorporates either heavy 4 x 4 lumber or 'built-up' two-by lumber nailed together to create posts and beams.

Framing members are placed far apart, which allows for wider openings — for windows and skylights. Floors and roofs are often made of 2-inch thick wooden planks connected with tongue and groove joints. Buildings made with post and beam framing typically have cathedral-type ceilings complete with exposed open beams. Generally, material and labor costs are lower than with the stud-and-joist method of construction.

Planning for a Room Addition

The steps involved in planning for a room addition are to check codes, draw up working plans, consult an architect, and secure building permits.

Read Up on Restrictions. Before you even begin to plan an add-on to your present home, find out about any possible restrictions. These can be numerous. Start out with local zoning. Zoning ordinances typically define sewage regulations and the use of municipal water. They also stipulate property setbacks, percentage of land coverage and heights of buildings. In some zones, such as historical districts, the style of architecture, type of materials, and even the color of paint may be regulated.

Look at your house deed. Check it for any restrictions regarding add-ons. You might be limited to only one story or a certain size. Also, as documented on this agreement, your use of certain materials might be prohibited.

Next, contact your local building department for a copy of their codes. These are usually an adaptation of national codes with appropriate changes to suit your locale. The design and construction of your project will have to conform to the guidelines established in this document.

Illustrate your Intentions. The next step is to commit your plan to paper in the form of a working

Post and Beam Construction. This barn, a reconstruction of an original, is made with timber frame construction, a special type of post and beam construction utilizing mortise and tenon joinery. Note the large supporting posts and the generous spacing between them. An attractive trademark of post and beam framing, when it is used in housing, is the interior's open beamwork.

drawing. You don't need to be a draftsman in order to do this, especially if your add-on is fairly small and uncomplicated. Most building departments will accept rough sketches. If your sketch is *very* rough, you can always hire a draftsman to rework it for you so that it's presentable.

As part of your plans, you should include what architects call a 'program' which is simply a statement of what each room will be used for and its dimensions. The drawings should consist of several views, and include doors, windows, stairways, closets, and heating and cooling ducts. Also indicate electrical and plumbing fixtures. If you want your drawings to look professional, use standard architectural symbols (page 75).

Appoint An Architect. This step can be skipped if your project is a small one or if you know exactly what you want and you're able to create drawings that are acceptable. On the contrary, if you don't trust your own ability to design a workable and attractive home addition, you should probably hire an architect at the outset instead of struggling with the design yourself.

Architects may charge an hourly fee or they may ask for a set fee which is usually about 10 percent of the estimated cost of the project. Experienced architects will be knowledgeable about both zoning restrictions and building permits, plus they will steer you toward a geologist if you're in an earthquake or slide area. What's more, the architect will be able to supply your local building department with all the drawings and everything that they need to issue you a building permit.

Proceed with a Permit. A building permit is required for almost any job imaginable that involves structural changes. Some of the projects that *don't* call for a permit are paneling, painting, and cabinetry — basic changes that won't 'alter the use' of the structure.

For starters, call on your local building inspector. This person will tell you whether you need a building permit and what is required in order to obtain one. Some of the information that's likely to be needed are the street address, lot, block, tract, or the assessor's parcel number of the location. Often too, you'll be asked to state the occupancy intent and the estimated value.

Depending on the complexity of your project, even more information might be required. Your building inspector should also inform you as to

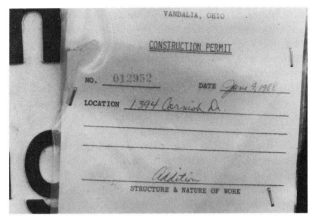

A Building Permit is required for any work that will alter the use of the basic structure. Keep it posted at your work site.

HOW TO STEP OFF A RAFTER

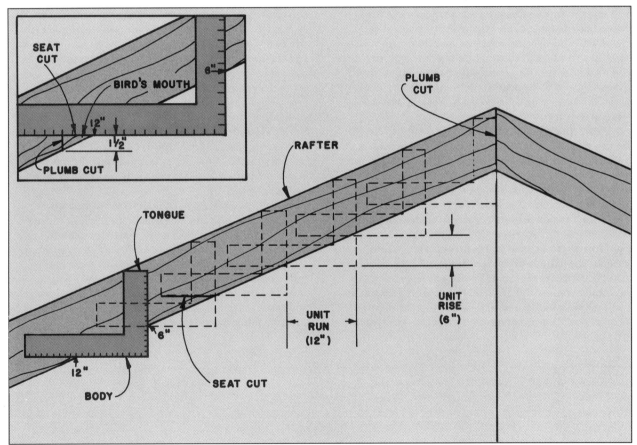

Choose a straight piece of lumber to make this pattern. Using unit rise and unit run, align your square as shown, with the body at the 12-inch increment and the tongue at the unit desired (shown is 6 inches). Draw a line along the tongue of the square. This will be the cutline for the plumb cut. Draw a line along the body of the square. Move the square to the point where the mark you have made inter-

sects the outside edge of the rafter. Continue marking and measuring in the same way. Do this a number of times to equal the unit run. For example, a seven-foot run is stepped seven times. Then, position the square one additional time and draw a line along the square's tongue. This mark indicates the plumb cut of the bird's mouth (see inset). To finish off the bird's mouth, reposition the

square to the original figures (in this case, 12 and 6) and then measure up from the rafter at the plumb cut a distance of 1½ inches for the depth of the top plate. If the rafter has no overhang, this plumb cut will extend the width of the rafter. If the rafter has an overhang, merely make a similar parallel mark for the tail cut.

whether you need additional electrical or plumbing permits. The permit is your formal permission to proceed with work so you should not begin until you have obtained it. Once it's in your hands, be sure to post the permit at your work site.

Putting Plans in Operation

For a simple and small add-on, you can probably plan on a weekend or a week of intensive labor, perhaps with the assistance of a helper or two. If it's been a while since you've done carpentry work or if you've never done it before, you're likely to find some muscles in your arms, legs, and neck that you didn't know you had. Aside from the physical exertion, you might experience a little mental anguish. If you run into a problem area or a question about construction, you will not have experience to guide you. When this happens, you'll need to research the answer or talk with knowledgeable people.

Knowing Your Own Limitations. There are several ways to analyze your project — by size, materials to be used, and difficulty of the skills involved. The size and scope of an add-on may dictate that you hire additional help just to get the framing done before the weather destroys your work. Materials and tools need to be hauled to the work site. Of course you'll also want to get the best possible price on your materials.

As for the difficulty factor of the skills, these will be rated low-level, medium-level, and difficult. As a general rule, low-level skills are within the domain of the do-it-yourselfer as are some medium-level skills. For the difficult skills, however, you might want to acquire professional help. Refer to the list below to determine which jobs you'll accomplish and which will be done by someone else.

Contractors and Subcontractors. When your project is a major add-on, consider hiring a general contractor to do all the difficult tasks. His role is to be totally responsible for the work and his asking fee will probably be between 10% and 15% of the total project cost. A good general contractor is invaluable since he'll be familiar with local building codes and he'll have the experience to negotiate the best prices for the various jobs.

Subcontractors, on the other hand, are contracted to do only part of the job. They may be employees of small businesses that specialize in masonry, drywall or the like or they may be merely 'independents'. In the case of independents, they will be *your* workers. You can pay them an hourly fee but don't forget that as their boss, you must also make sure that they have their own worker's compensation insurance. If the work goes on for an extended period of time, you'll also be obliged to register with state and federal governments as an employer. This means withholding income taxes and disability insurance, paying social security and unemployment costs.

Your responsibilities when working with subcontractors are to schedule and coordinate all the work, make appointments for permits and inspections, oversee the ordering and delivery of materials and write the checks.

As you can see, you have several options — do all the work yourself, or hire a general contractor or subcontractors to do all the difficult tasks while you do the rest yourself. Or you could be an assistant to your contractor or subcontractor if they agree to work in this way.

Before hiring anyone to work on your home in any degree, ask them for references and licenses. If you can, actually inspect their work. Check with the Better Business Bureau to see if any complaints have been registered against them. Check with former clients to make sure they were satisfied. This is the only sure way to know exactly what kind of workmanship to expect.

When you've made the decisions about who is to do what, then the next step is to spell it all out in a contract. The contract should relate very specifically how and when the work is to be completed, by whom, and with what materials. If your project is a major one, be sure to have an attorney look over your contract for any loopholes. This might be a wise idea in any case no matter what the size of the project unless you're willing to take the risk.

Low-Level Skills	Medium-Level Skills	Difficult Skills
☐ Gutters and downspouts	☐ Cabinets and countertops	☐ Foundations
☐ Floor coverings	☐ Roughing in plumbing	☐ Masonry
☐ Interior doors and trim	☐ Plumbing fixtures	☐ Heating and cooling systems
☐ Wall coverings	☐ Wiring	☐ 'Complicated' framing
☐ Painting (interior and exterior)	☐ Electrical fixtures	☐ Drywall
☐ Clean-up	☐ Exterior sheathing	
☐ Landscaping	☐ Exterior siding	
	☐ Windows and doors	
	☐ 'Easy' framing	
	☐ Garage door and opener	
	☐ Floor and wall tile	
	☐ Roofing	

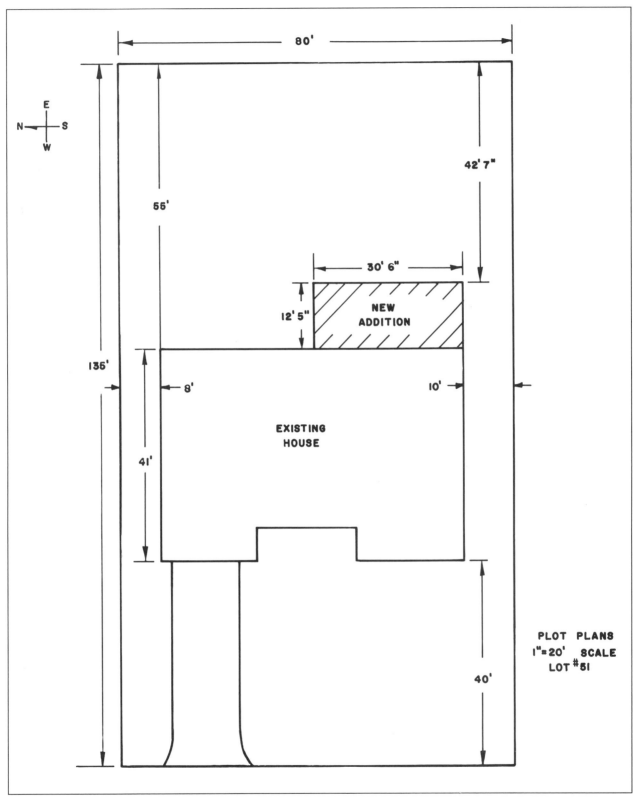

Shown here is a typical *plot plan* indicating property lines, building lines, the existing structure and the new addition. An *elevation drawing* (page 74) shows views of the existing house with the new addition. When you make rough drawings, use standard *architectural symbols* like those shown on page 75 to indicate heating and cooling systems, electrical components, plumbing fixtures, etc.

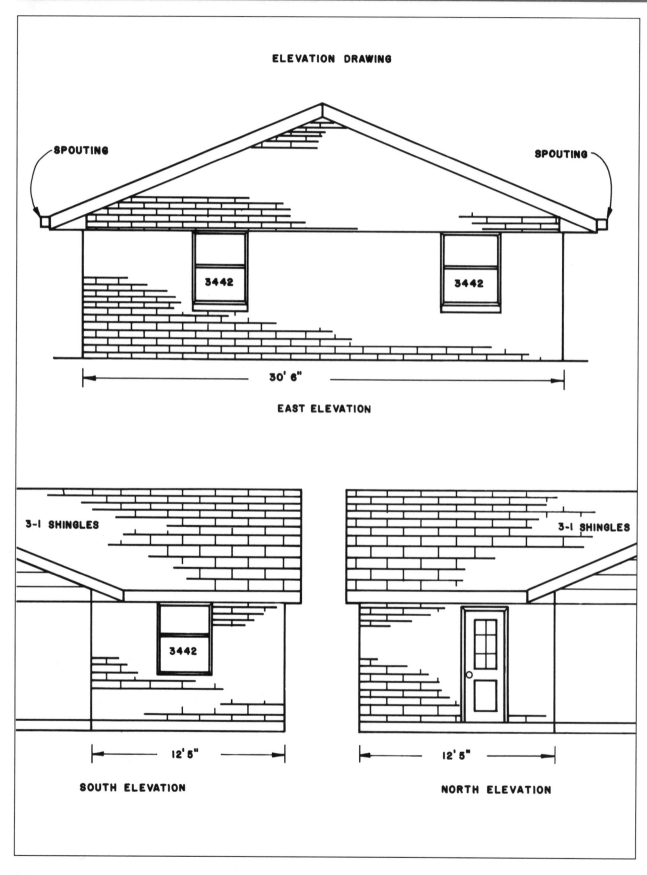

ELEVATION DRAWING

SPOUTING

SPOUTING

3442

3442

30' 6"

EAST ELEVATION

3-1 SHINGLES

3-1 SHINGLES

3442

12' 5"

12' 5"

SOUTH ELEVATION

NORTH ELEVATION

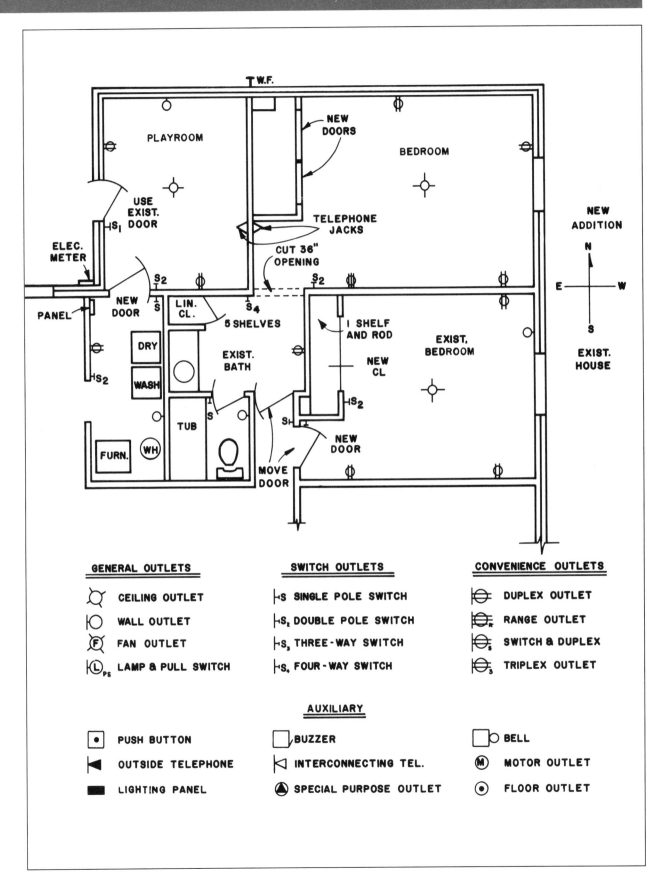

GENERAL OUTLETS

- ⊗ CEILING OUTLET
- ⊙ WALL OUTLET
- Ⓕ FAN OUTLET
- ⊙ₚₛ LAMP & PULL SWITCH

SWITCH OUTLETS

- ⊣S SINGLE POLE SWITCH
- ⊣S₂ DOUBLE POLE SWITCH
- ⊣S₃ THREE-WAY SWITCH
- ⊣S₄ FOUR-WAY SWITCH

CONVENIENCE OUTLETS

- ⊖ DUPLEX OUTLET
- ⊖ᵣ RANGE OUTLET
- ⊖ₛ SWITCH & DUPLEX
- ⊖₃ TRIPLEX OUTLET

AUXILIARY

- ▪ PUSH BUTTON
- ◀ OUTSIDE TELEPHONE
- ▬ LIGHTING PANEL
- ☐ BUZZER
- ◁ INTERCONNECTING TEL.
- ⬤ SPECIAL PURPOSE OUTLET
- ☐ BELL
- Ⓜ MOTOR OUTLET
- ⊙ FLOOR OUTLET

6

The Fundamentals of Framing

Now you're ready to work from the foundation to the roof-top and learn the fine points of framing. As you read this chapter you'll learn how to build the framework for your add-on.

Floor framing is outlined; you'll learn how to give your structure a sturdy platform — one that's level and problem-free. Next you'll see wall framing which is actually anchored to the floor. You'll discover how to 'turn corners' and create openings for windows and doors. Although the subject of roofs can be quite complex, here you'll see the basic roof styles and ways of constructing them.

The idea of building an addition onto your house might be overwhelming but it doesn't have to be, as explained here. You're shown how to tie in a wall to an existing structure and also how to connect a shed roof and a gable roof to existing roofs.

There are many products that can make your job simpler, such as metal connecting plates and pre-cut roof trusses. While building, continually check to see if your work is level and true. Finally, use 'safety sense' and follow the safety rules presented at the beginning of this book.

Framing the Floor

Once the footers and foundation walls are constructed, it's time to start the framing process. This work will be done with a more pliable and forgiving material — wood.

If you've chosen to have a slab foundation, then there is no floor framing to be done. All you have to do is attach the mudsills and then move on to building and erecting the walls. But if you have a subflooring type foundation, you'll need to frame the floor in these four stages:

[1] Attach mudsills on top of the foundation walls.
[2] Install girders.
[3] Attach joist headers, joists, and metal or wooden bridging between the joists.
[4] Lay the plywood subflooring.

Mudsills — Straight and Sealed. Mudsills are extremely important components of your structure. When choosing materials for your mudsills, usually 2 x 6s, look for the longest and straightest boards you can find. When installing mudsills, pay particular attention to detail. The smallest error multiplies as the structure grows.

Even if you've had your footers and foundation walls put in by a professional, the first thing you must do is check the foundation for square. Do this by running mason's twine from opposite corners of the foundation walls to create a large X. The length of the diagonals should match. If they don't, then mark the bad corner (or corners) with the distance that it was 'off'; for example, 'long ¼ inch'. This establishes the good corners and allows you to make up for discrepancies in the foundation by 'truing up' the mudsills.

Measure in from each corner the width of the mudsill material (usually 5½ inches) adding or subtracting these 'long' or 'short' corners. Snap chalk lines between the marks for aligning the inside edges of each mudsill.

Next, drill the holes for the anchor bolts and install the mudsills using the instructions on the right. Check your local building code to verify the need for a sill sealer and metal termite barrier. Make any adjustments for level by driving wooden shims, usually cedar, under the mudsill before the sill anchors are tightened.

Midspan Supports — Posts and Girders. Depending on the size of your add-on, you might need to add posts and girders. If it measures wider than 16 feet, this will be necessary. As always, be sure to check your local building code for the exact specifications.

In any case, the span of the girder, its depth, and its thickness are all related to each other. Other factors are the size of the posts and the load of the structure — i.e., will it be one-story or two-story? All this, of course, was determined when the structure was designed so that the proper allowances were made.

To support the girder, a pier or several piers will be built of concrete, or a post anchor will be embedded in the footing. Special pockets or notches

will be made in the foundation walls to support the ends of the girders.

As for materials, the posts themselves can be made of solid or built-up lumber, or they can be composed of concrete-filled steel columns (called *lally tubes*). The girders, too, may be of several kinds of materials — steel, solid lumber or built-up lumber. In many areas, wooden posts and girders are required to be made of decay-resistant or pressure-treated lumber.

Typically, a built-up wooden girder consists of three thicknesses of 2-by dimension lumber nailed together with 20d nails. Nails should be spaced at top and bottom edges, about 32 inches apart.

Position the posts first and then lay the girder on top. Begin by measuring from mudsill to mudsill where the center of the girder will be placed. Check for level and then measure to the post anchors or piers. This will give you the needed lengths of the posts. Brace up the posts and check carefully for level and plumb. Use special metal plates to fasten the posts at the bottom and to attach the girder to the post.

You'll need several helpers to fit a large girder into its foundation wall pockets. If it isn't perfectly level, use metal shims to make it right. Finally, toe-

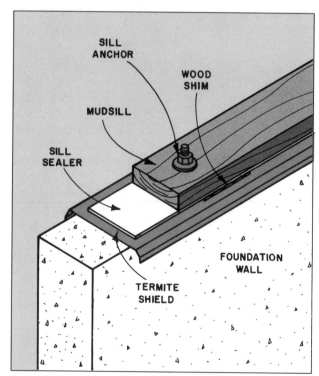

Installing the Mudsill. Use lumber specified by your local building code for the mudsill — usually decay-resistant or pressure-treated wood. Hold the mudsill next to the foundation wall and transfer the positions of the sill anchors onto it. Drill holes slightly larger than the anchors. If your local building code calls for a sill sealer, place it between the mudsill and the foundation wall to act as a vapor barrier. If termites are a problem in your area, also add a termite shield. Screw down the sill anchors by hand; then check the mudsills for level. If needed, add wooden shims.

nail girders into the posts with 10d nails, or, if metal 'post caps' are being used, screw or nail them in.

The Floor Joist System. One of the most time-consuming steps of framing the floor will be installing the joists, the horizontal members that stretch from mudsill to mudsill or from mudsill to girder.

Special Floor Framing. At this stage, you will need to consult your plans since floor openings such as those for stairways and plumbing fixtures need to be reinforced. Special framing members are used — *headers, trimmers,* and *tail joists.* Headers and trimmers actually frame the opening, so they are usually doubled. The tail joists are the 'short joists' that continue on where an opening disturbs the normal joist pattern.

Use double joists around the edge of a bathtub, especially a heavy one, and double headers to frame plumbing. Other places where this is necessary are below interior partition walls and around floor openings, such as those for stairways.

For an opening more than 4 feet wide, use double headers. When headers will be longer than 6 feet, and tail joists will be longer than 12 feet, use joist hangers. Special floor framing instructions are shown on this page.

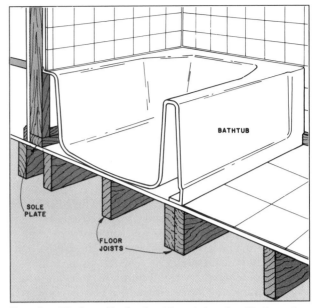

Framing Under a Bathtub. Floor joists should always be positioned with the plumbing in mind. Single joists running parallel to the wall are separated to allow for pipe runs. Double joists on the outside edge of the bathtub provide extra support for this heavy fixture.

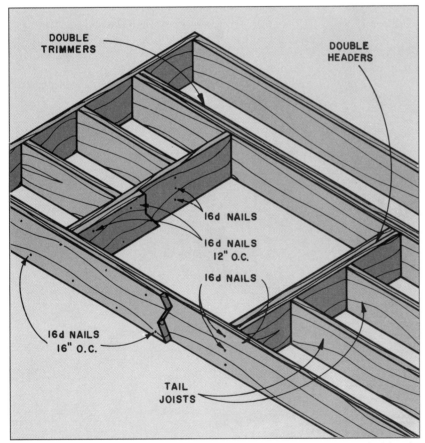

Framing for a Floor Opening. Begin by installing a trimmer on each side of the opening. Next, attach initial headers to the trimmers with three 16d nails. Cut tail joists to length and fasten them with 16d nails. Add second, inside headers, nailing them to trimmers and then to initial headers. Finally, add the second trimmers, attaching them along top and bottom edges of the first ones with 16d nails.

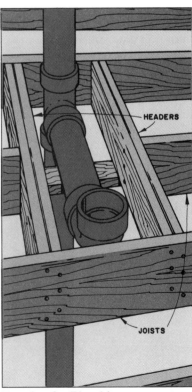

Framing for Plumbing. Special provisions might be necessary when running drain and vent pipes and other plumbing through the joists. Depending on varying local codes, a joist may be cut into for a pipe only at its end quarter. Once you have done this, reinforce the cut by installing headers on both sides. Also, brace the pipe between the double headers.

Choosing Joist Materials. Your choice of materials for floor joists will be based on many factors. Among them are: the load, the length of the span, and the space between the joists. There are several species of lumber to choose from — pine, hemlock, cedar, and Douglas fir. Generally grade 2 or 3 is used. The sizes of joists commonly used are: 2 x 8, 2 x 10, and 2 x 12.

All of these elements depend on each other. For example: If No. 3 grade Douglas fir 2 x 8 joists are placed 16 inches on center (O.C.), they may span a maximum of 10 feet, 7 inches. If the grade, species, size of joists or space between joists differs, then the maximum allowable span will change. The easiest way to determine your specifications is to check your local building code. Give the local agency the details of your plan and they will recommend the appropriate material and guidelines. When purchasing wood, get the driest, straightest lumber you can find.

Joist Layout. Beginning at the *stringer,* the first joist, mark the mudsill with X's at appropriate intervals. If your joists will be 16 inches O.C., make straight marks at 14½ and 17½ inches; then draw the X in the space. If joists are to be doubled, continue to mark for 16 inches O.C. and then fill in the marks for the additional joists when you're through marking the length of the mudsill. The last joist may fall short of the 16-inch spacing; include it anyway.

When you measure the opposite mudsill for joists, use the same method unless you're overlapping joists at a central girder. In this case, allow 1½ inches and begin the first marks at 13 and 16 inches. Then proceed to measure and mark as usual.

Fastening Floor Joists. Use the technique below to install the joists. If you aren't very experienced at toenailing, another method of installing the joist header is to first end-nail several joists to it and then toenail it to the mudsill. The weight of the joists will help to secure the joist header. You might also find it easier to use joist hangers to attach the joists to the headers.

Reinforcing the Floor. When the floor joists, headers, and trimmers are all laid, you should next reinforce them with bridging or blocking (right). Your building code will specify the length of span that needs to be bridged, plus it might recommend what type to use. Although bridging does not provide critical support, it apparently does prevent the joists from rotating at their midspans.

A trick to installing blocking is to stagger successive members slightly to leave room for nailing. Metal bridging requires no nails. Simply hammer in the prongs to the top edge of the joists.

Adding the Subflooring. The last step in the sequence of floor framing is to add the subflooring. The subflooring is fastened directly to the floor joists. Once the subflooring is in place, you'll use it as a work surface when assembling the wall sections.

Subflooring Materials. In the past, subflooring was usually made of boards placed either straight or diagonally across the floor joists (right). This method is still used, although it is most commonly seen today with post and beam structures.

A most popular modern material is plywood; it's strong and it's also inexpensive compared to board flooring. Typical thicknesses of plywood to use are 7/16-inch and 3/4-inch; use 3/4-inch when joists are 24 inches O.C. Grades that are good for subflooring are Sheathing and C-Plugged.

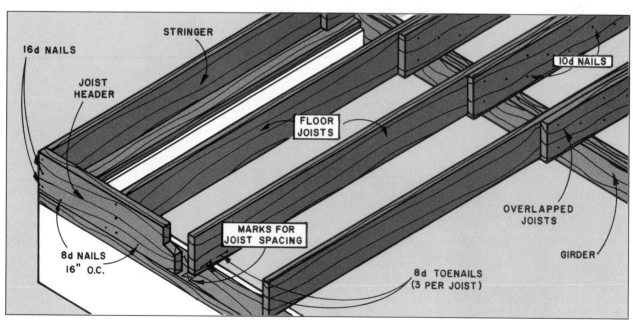

Installing the Floor Joists. First, install the joist header to the mudsill with 8d nails positioned every 16 inches on center. Then nail each joist end to the joist header with three evenly paced 16d nails. Finally, toenail each joist end into the mudsill, as shown, with three 8d nails. Where joists overlap at a central girder (a minimum of 4 inches) they should be fastened with 10d nails and toenailed to the girder.

THE FUNDAMENTALS OF FRAMING

Bridging of floor joists is required by most building codes. Joists should be bridged every 8 feet with wooden blocks (shown on the left) or with metal diagonal bridging which usually requires no nails (shown on the right).

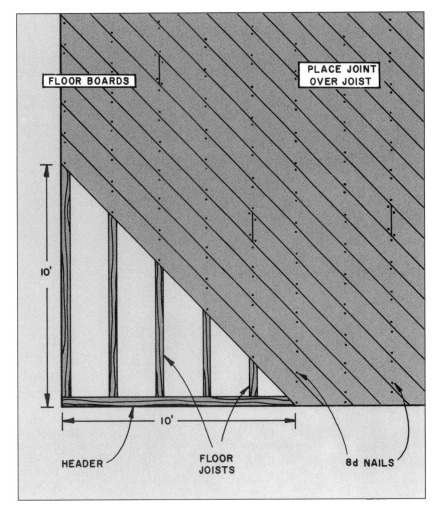

Laying Board Subflooring.
Measure for and mark points at 10 feet along the header and 10 feet along the end joist. Align the first board to these points and nail securely. Work inward away from the corner leaving slight gaps for weathering while the building is in progress. Fasten boards to all members with two 8d nails. Then fill in the corners.

THE FUNDAMENTALS OF FRAMING

Laying the Panels. When you install the plywood, use the diagram below to lay out the panels correctly. Stagger them and seam the ends directly over the centers of the joists. If the panel spans across overlapped joists, such as where they meet above a girder, nail a wood block to the offset joist and then begin your new row of panels.

For an even stronger floor, in addition to nails, use glue. Choose a special elastomeric construction adhesive and apply a bead to each joist. Also apply glue to the panels. When this method is used, fewer nails are needed.

Vertical Expansion — Framing the Walls

When the subflooring is completed, or in the case of slab foundations, when the mudsills are in place, you can begin the wall framing. The process involves assembling whole sections of the walls and then raising them into place on the mudsills. This should be done in the following order:

1. Construct and raise the exterior side walls.
2. Construct, raise, and tie in the end walls.
3. Construct, raise, and tie in the interior partition walls to the exterior walls.

Before you begin, you'll need to understand clearly what kind of wall you're working with.

Walls — Definitions and Design Tips. Walls can be built differently depending on where they are positioned in the structure. Exterior walls give protection from the elements so they must be constructed in a certain way. The walls that bear loads, called *bearing walls* must be strong. *Nonbearing walls* can be generally 'less solid' than bearing walls. To distinguish between these wall types, see the diagram on page 83.

In the past, the customary way of framing exterior walls was to use 2 x 4s placed at 16-inch intervals. This is still being done, but an increasingly popular method is to use 2 x 6s placed on 24-inch centers. This provides more space for better insulation — an important economic factor. Check your building code for the guidelines in your area. Regardless of which method you choose, the framing process is basically the same.

Whichever size studs you use, you'll probably find that your lumberyard lists 92¼ inches as the standard stud height. This dimension, which equals 7 feet 8¼ inches might be puzzling when you consider that standard ceiling height is 8 feet and wall coverings such as drywall sheets are available in 8-feet lengths. The reason for this seemingly odd dimension is that allowances are made for the top, cap and sole plates.

Partitions, the interior nonbearing walls, are usually made of 2 x 4s placed only 24 inches on

LAYING THE SUBFLOORING

1 If you're using an adhesive and nails to secure the plywood, run beads of the adhesive across the panels and on every joist to be covered.

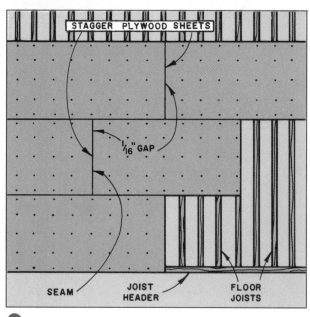

2 Lay 4 x 8 foot sheets perpendicular to the joists. Stagger them so every other row begins with a half sheet. Align sheets leaving a ¹⁄₁₆-inch space at the end and ¹⁄₁₆-inch gaps at each seam. Snap chalk lines to find joist centers.

center. The framing around wall openings in these walls can be of lighter lumber as well.

Basic Wall Assembly. The framing for any wall includes evenly spaced studs placed between a sole plate and a top plate. When walls have openings for doors and windows, extra framing members are required. Frame the individual wall sections in the order described on pages 84 and 85. A wind brace may be used on the walls. There are several types, including a K brace, and a wooden let-in brace. Some builders use metal braces that are wedged into slots sawn in the studs. You might need to add fireblocks — pieces of wood — between the studs. Check your local building codes for information on bracing and fireblocks.

Making Spaces for Doors and Windows. If your new walls are to have doors or windows, you'll need to create openings for them in your framing. Refer to the illustrations on pages 86 and 87 as you study the following information.

The first thing you need to know is the *rough opening* of the window or door unit as specified by the manufacturer. If this information is not given, measure the unit and then add ½ inch at the top and sides for level and plumb adjustments. Locate the center point for your opening and then measure off half of the rough opening dimension in both directions. These are the locations for your trimmer studs.

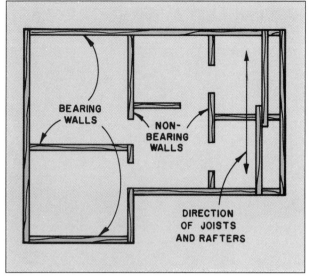

Bearing and Nonbearing Walls. Bearing walls support the weight, or load, of the structure; nonbearing walls do not. Joists are always run to the width, not the length of the structure and the bearing walls support the ends of the joists. Bearing walls also support the joists at a midspan, over an interior foundation wall or a girder. All other walls are nonbearing. Design requirements for interior nonbearing walls, known as *partitions*, are more lenient than for bearing walls.

3 With a sledgehammer, drive sheets together.

4 Use 6d nails for material up to ½ inch thick; 8d nails for thicker sheets. Space nails 6 inches apart at panel ends; 10 inches apart at intermediate joists. When using glue, space nails 12 inches apart.

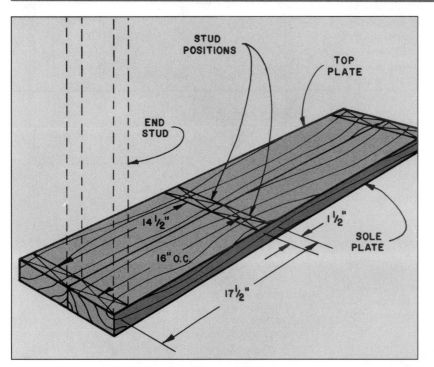

1 Begin by cutting the top and sole plates to length. If you use more than one piece of lumber for each plate, measure each piece to meet stud centers. Lay the top and sole plates together with faces up and mark for stud positions. Measure 16 inches on center and then draw rectangles with Xs to show the center points and exact stud positions.

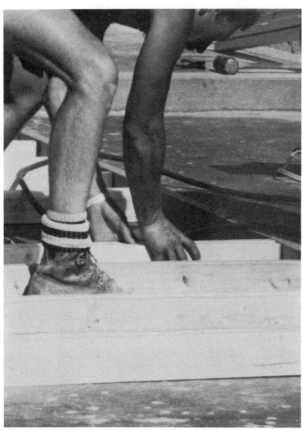

3 Hammer and fasten each stud with two 16d nails. Position fire-blocks 4 feet above the sole plate or as required by code; stagger them slightly for easy nailing.

4 If necessary, add a wind brace to the exterior side of the framing. Using the brace itself as a guide, make pencil marks for the grooves in which the brace will be secured.

Lay out the cap and sole plates with the Xs facing to the inside.

Cut the grooves with a circular saw. Then, nail the wind brace in position.

DETAILS OF DOOR FRAMING

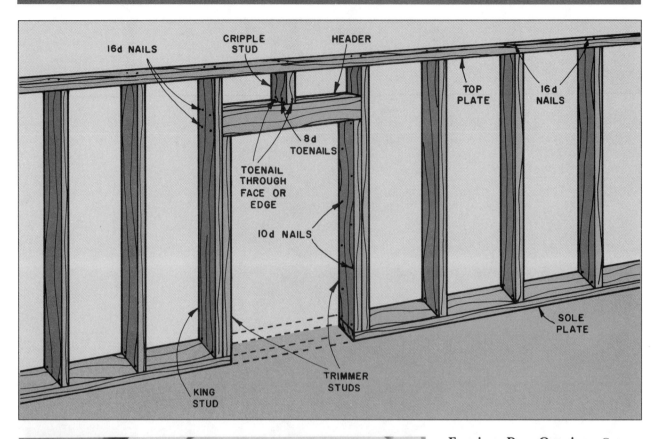

Framing a Door Opening. Cut and install the king studs and nail them to the top and sole plates with 16d nails. Next, cut and nail trimmer studs to the king studs; use staggered 10d nails placed 16 inches on center. Position the correct size header on top of the trimmers and secure it with 16d nails through the king studs. Measure and cut cripple stud(s); attach them from the top plate, 16 inches apart, with 16d nails. Also toenail them with 8d nails into the header.

DETAILS OF WINDOW FRAMING

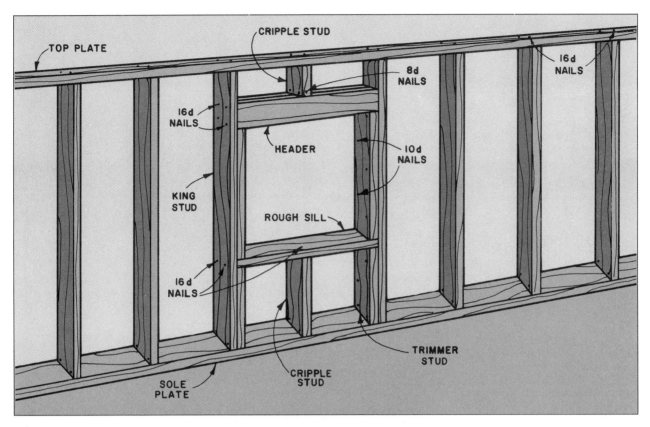

Framing a Window Opening.

Install the windows so that their tops are flush with the tops of the door openings, or, in other words, the bottom edges of the headers match. This dimension is usually 6 feet, 8 inches. Begin by measuring the height from the top of the sole plate to the bottom of any door header already installed. Mark this height on the king studs of the window opening. From this measurement, subtract the height of the manufacturer's rough opening; the result is the mark for the top of the rough sill. Subtract another 1½ inches for the thickness of the sill and make an additional mark. From here to the top of the sole plate is the dimension needed for the length of the lower cripple stud(s).

Cut and install the king studs the same as for a door opening on the preceding page. Measure for and cut the lower cripple stud(s). Next, cut the rough sill and fasten it to the king studs with 16d nails. Insert the cripple stud(s) and fasten them to the sill with 16d nails. Add the trimmer studs and the upper cripple stud(s) same as for a door opening.

1 Shown are constructed walls ready to be raised. Arrange all the walls in their proper positions, as close as possible to the joist headers or other appropriate framing members to which they will be attached.

2 Use at least three people for this job. Begin by positioning the wall with its outside studs aligned parallel to the structure. Nearby, lay out 1 x 4 braces. All together, pick up the wall by its top plate, grip it underneath, and walk it to an upright position.

3 Check the wall for plumb.

4 Nail temporary diagonal braces at the tops of the wall studs and anchor them to 2 x 4 blocks nailed to the subflooring or secured on the outside of the structure to stakes driven into the ground. Check the wall for plumb. When the other walls are raised and plumbed, check the diagonals corner to corner. Fasten their sole plates through the subfloor and into the joists and joist headers. Secure the sole plates down by nailing two 4-inch nails between each pair of studs; the nails must reach a joist or the joist header. Plumb the corner studs and secure them with diagonal bracing. Do not nail plates beneath doorways. If you have a slab type foundation, drill holes in the sole plates and secure them with the sill anchors.

The trimmer studs are faced to the supporting *king studs*. In your planning, try to position the opening so that at least one existing (evenly spaced) stud will be used as a king stud.

The headers that frame the top of the door and window openings are usually composed of 2-by lumber turned on edge and faced together with a piece of ½-inch plywood sandwiched between them. An optional material is 4-by lumber. Use the chart on the right to determine correct header size for a one-story structure, and then verify it with your local building code. If there will be two stories, always choose the next largest header size.

A variation in header size is necessary when you're using 2 x 6 studs for the wall framing. In this case, you can use one solid 6-by beam. Another variation is in the framing of a partition wall opening; these require only a single 2 x 4 or 2 x 6 header.

The formula for building headers is to cut the 2-bys and the plywood to the length between the king studs. Fasten these components with 16d nails spaced 16 inches apart at top and bottom edges.

DETERMINING HEADER SIZE

Width of Opening	Size of Header
Up to 4 feet	4 x 4 two 2 x 4s on edge
4 feet to 6 feet	4 x 6 two 2 x 6s on edge
6 feet to 8 feet	4 x 8 two 2 x 8s on edge
8 feet to 10 feet	4 x 10 two 2 x 10s on edge
10 feet to 12 feet	4 x 12 two 2 x 12s on edge

Finishing Up. When the exterior wall sections are assembled, raise them (page 88) and then tie them together at the corners (below). Every step of the way, check for level and plumb.

FRAMING CORNERS

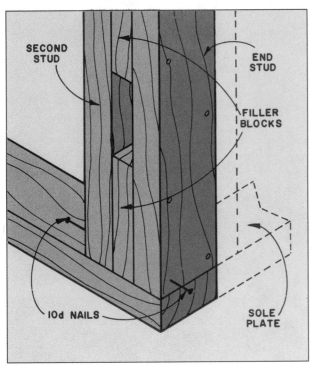

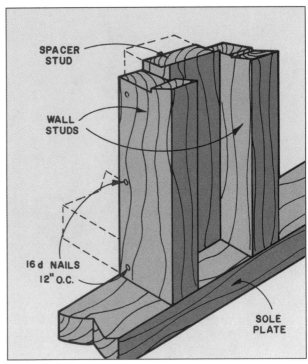

Exterior Corners. After framing and raising the long exterior wall, cut and nail filler blocks, short 2 x 4s, to the end stud. Then add a second stud toward the inside; also fasten it to the filler blocks. Nail both studs with 10d nails. The resulting surface will create a base for fastening the abutting exterior wall plus the interior wall materials.

Interior Corners. Where an interior wall ties in, three studs must be used. The two additional studs may be added after the interior wall is raised. Use 16d nails to fasten the outside studs to the spacer stud. Note the position of the spacer stud which provides a base for the end stud of the interior wall.

THE FUNDAMENTALS OF FRAMING

STRAIGHTENING THE TOP

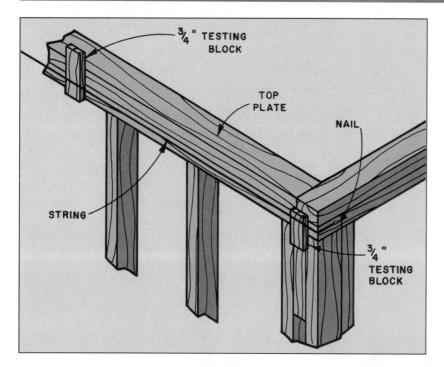

¾" TESTING BLOCK

TOP PLATE

NAIL

STRING

¾" TESTING BLOCK

1 To straighten the top, run a tight string along the top plate, tying it to nails around the ends. Block the string out at the ends with a small ¾-inch thick block.

2 Use a ¾-inch thick test block and check at various points to make sure the string is the correct distance from the plate.

3 Push or pull the top of the wall in or out to correct any points not in line. Brace the wall by nailing one end of a board to the stud at the top and the other end to a block fastened to the subflooring.

THE FUNDAMENTALS OF FRAMING

Next you need to straighten the top of the exterior walls. This is accomplished by running a string along the top plate and moving the wall in or out to make it measure an equal distance from the string (page 90).

When the exterior walls are anchored, install the interior walls. These are constructed and raised just like the exterior walls. Finally, install the cap plates (right). Leave all braces in place until the ceiling joists and rafters are installed.

Tying into an Existing Wall. To tie in a new wall to an existing wall, you'll need to follow these five steps:

1. Determine the starting point.
2. Cut into the existing wall.
3. Cut into the existing roof.
4. Brace the existing wall.
5. Install new wall studs.

This process, which is shown on pages 92 and 93, should be done with accurate measuring and frequent checks of level and plumb. **Special care should be taken during the cutting steps since you'll be holding your saw at a very awkward angle.**

Framing the Roof

With the floor and walls framed, you'll probably want to get a roof on as soon as possible to protect the work you've done so far. Don't rush through this step, however. Precision is more important than ever, and in the actual framing process, safety practices are crucial.

Here you will learn how to frame the basic gable roof by two different methods — with trusses and with a ridgeboard and rafters. If you're building a relatively large structure such as a garage, trusses are recommended. If you're making a small add-on, use the second method.

Before you get started, review the section on roofing in Chapter 5, pages 67-71. Next, check the walls of your structure for square by measuring diagonals across the cap plates.

Using Roof Trusses. Factory made trusses are widely used in home construction. They permit fast, easy roofing and the step of adding braces is eliminated. Moreover, a truss includes a horizontal member that takes the place of the ceiling joist. In recent years, trusses have become more popular — especially the type made with metal connectors.

Three popular types of trusses are shown on page 92. Study this and then check to see what's available at your lumberyard. Truss design takes into account the load and slope of the roof. Make sure that your plan is approved by your local building department. They will have special tables that match up roof slope with appropriate building materials.

There are several methods for mounting trusses (pages 94 and 95). They may be inverted and hoisted up by hand one-by-one. Special long poles may be used by people on the ground to maneuver and posi-

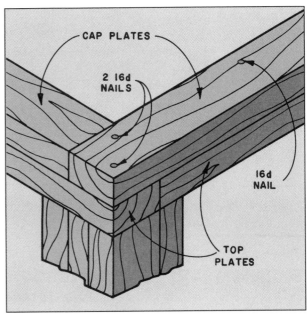

Fastening the Cap Plates. Fasten the cap plates onto the top plates. They should measure at least 4 feet and offset the joints in the top plates. Use 16d nails spaced 16 inches apart; secure all joints with two nails. Overlap corners and intersections as shown.

tion the trusses. Perhaps the most common method is to use heavy machinery to lift the trusses up and then have three men install them. One positions and raises the truss while the other two tie it in to the cap plates.

Using a Ridgeboard and Rafters. When you use this traditional method of building a gable roof, follow these steps:

1. Attach ceiling joists to the cap plates.
2. Install the end rafters and ridgeboard.
3. Attach remaining rafters.

The function of *ceiling joists* is two-fold. They support the ceiling materials below and they help to brace the exterior walls of the structure from the weight of the rafters above. They are usually lighter than their counterpart floor joists below.

Ceiling joists can be made of 2 x 4, 2 x 6, or 2 x 8 lumber. They can be placed 16 or 24 inches apart, usually spanning the short dimension of the structure. Like all framing members, the kind and size of wood to use is dependent on many factors. Check your local building code for a span chart and a recommendation of materials.

To lay out ceiling joists, simply pencil their positions on the cap plate. If rafters will be spaced the same distance, then place them next to each other. If they're spaced differently, at 16-inch and 24-inch intervals, lay them out to meet every 48 inches.

Unlike floor joists, ceiling joists cannot be overlapped. When you need to begin a new joist, such as over an interior bearing wall or over a beam, butt two joists together. Toenail them to the top plate and then nail 2-by 'bearing' blocks to the butt joint for extra support.

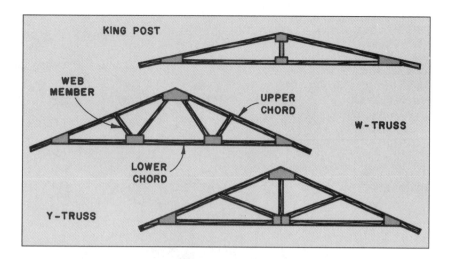

Truss Designs. Shown top to bottom are three simple types of trusses commonly used in residential construction. The simplest is the *king-post* which utilizes one central vertical post. The *W-truss* and *Y-truss* are also economical and easy to use. Framing members at the top are called *upper chords.* The bottom ones which serve as the ceiling joists, are called *lower chords.* Interior components are referred to as *web members.*

TYING INTO AN EXISTING WALL

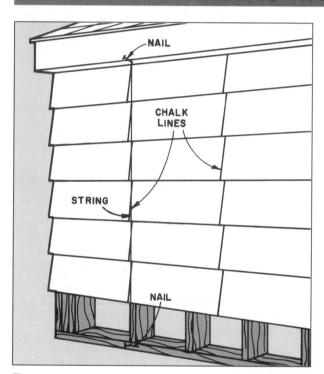

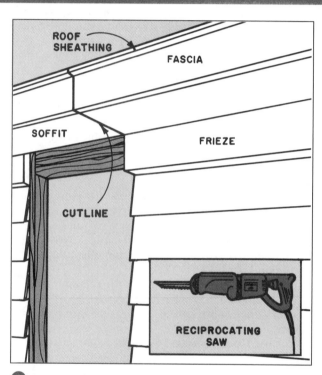

1 If the new wall meets an exterior corner, snap a chalk line about 10 inches from the corner. If the new wall meets the middle of an existing wall, as shown, snap two lines — one ⅝ inch outside the edge of the new foundation, the other approximately 12 inches in from the first. This will create an opening between two studs. Repeat the process on the other side of the foundation. Make horizontal marks at the height of the interior room's ceiling. Cut through the siding on the marks with a circular saw fitted with a carbide-tipped blade. Create two openings by prying off the siding panels, working from top to bottom.

2 To remove an overhang, you'll need a *reciprocating saw,* a powerful saw that's ideal for cutting into walls. You may be able to rent this tool; if so, make sure that it's fitted with a 12-inch blade for cutting into walls with nails. Begin by removing part of the outer roofing material to expose the roof sheathing. Then extend, with a square, the outside lines made in step 1. Mark the frieze, soffit, fascia, and roof sheathing. With the reciprocating saw, carefully cut into the frieze at both cutlines and pry away the material; *do not cut into the top or cap plates.* Continue to cut away and remove all the soffit and fascia material above the new foundation.

THE FUNDAMENTALS OF FRAMING

Two people should work on ladders at opposite ends to install ceiling joists. Toenail the joists to the cap plates with three 8d nails. If a cap plate of a partition wall is located parallel to and between two ceiling joists, add horizontal nailing blocks between the joists. Toenail the blocks to the plate. Frame any ceiling openings by the same method that floor openings are framed (page 79).

To install a *ridgeboard* and the *rafters,* first read the material on pages 68-71 regarding slopes and stepping off rafters. Create your rafter template using the instructions on page 101 and cut all the rafters for the roof. Then follow the instructions on pages 96 and 97 to complete the process.

Framing the Gable Ends. Whether you use trusses, or a ridgeboard and rafters, you will need to frame the gable ends. This involves putting equally spaced gable studs between the end rafters and cap plate.

Measure and cut the gable studs by first installing a single, central stud. Next to it, trace the outline for a stud while checking it for plumb. Then measure for the stud next to it. The difference between the length of these two studs is called the *common difference.* All studs will vary in height equally by this dimension. Also, the rafter angle on the first stud will be identical on all the rest.

Use the instructions on page 98 to do the framing. If you wish to have a gable overhang, build it out by using barge rafters and outriggers. Be cautious when doing work here though; even with scaffolds, it's a difficult space in which to maneuver.

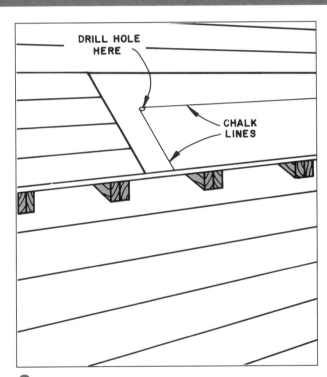

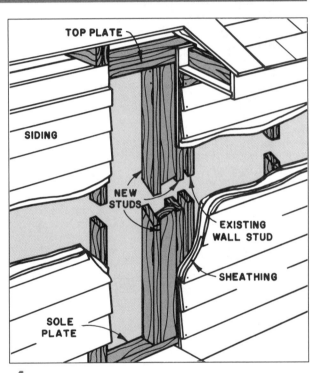

3 From underneath, drill pilot holes through the roof sheathing. The holes should mark the precise spot where the outer face of the new wall meets the outer edge of the existing cap plate. Snap a chalk line between the two holes. Also, snap short perpendicular chalk lines as shown. Cut the roof sheathing on the cutlines; pry it off. Use a level to make vertical marks on each exposed rafter. Cut away the rafters with a reciprocating saw.

4 Brace the cutouts by inserting studs between the sole and top plates. If you meet any existing wall studs, do not remove them; simply nail the new studs to them. Align the first stud flush to the house siding. Toenail it to the plates and then fasten it to the siding with nails spaced 6 inches apart. Place the second stud so its face is flush with the plates and the outside edge of the first stud; toenail it to the plates. Position the third stud, as shown; toenail it to the plates, and then face-nail it to the second stud.

CAUTION Always wear proper eye protection when using power tools (pages 3 and 4). Also, be sure to keep your hands out of the danger zone (page 5).

1 Begin by marking the position for each truss on the cap plate. Use at least three people for this job. Have all the trusses that can be safety lifted by one man positioned on top of the structure. The middle man raises up the truss.

CAUTION

Raising trusses involves working at high elevations. Make sure that the 'end men' have good balance and steady footing.

3 Make periodic checks for even spacing at the peak of the truss. These workers are installing a temporary horizontal brace at the ridge line of the roof to help keep the trusses aligned until the roof sheathing is applied.

THE FUNDAMENTALS OF FRAMING

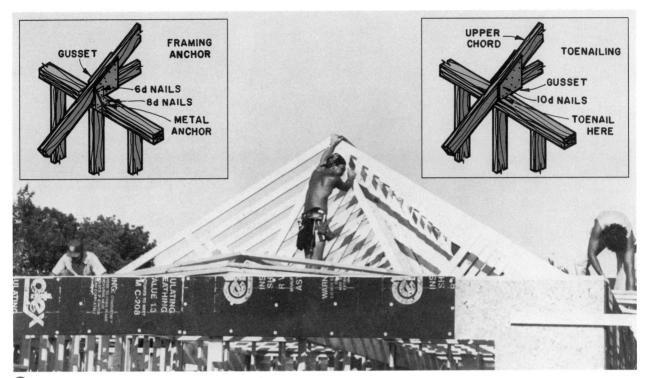

2 Next, the two end helpers align the lower chords of the truss with the marks on the cap plates. Fasten them by toenailing. If desired, use framing anchors or metal plates for reinforcement. (The insets show two nailing methods.) Repeat the entire process with each truss.

4 Heavy machinery is used to install the end or final truss which has been preassembled with sheathing and barge rafters.

5 Once the roof trusses have been erected, you can check the lower chords for alignment. This worker is using a board to check for equal spacing between the trusses. After any irregularities have been corrected, the lower chord is nailed to a 2 x 2 horizontal supporting member.

THE FUNDAMENTALS OF FRAMING

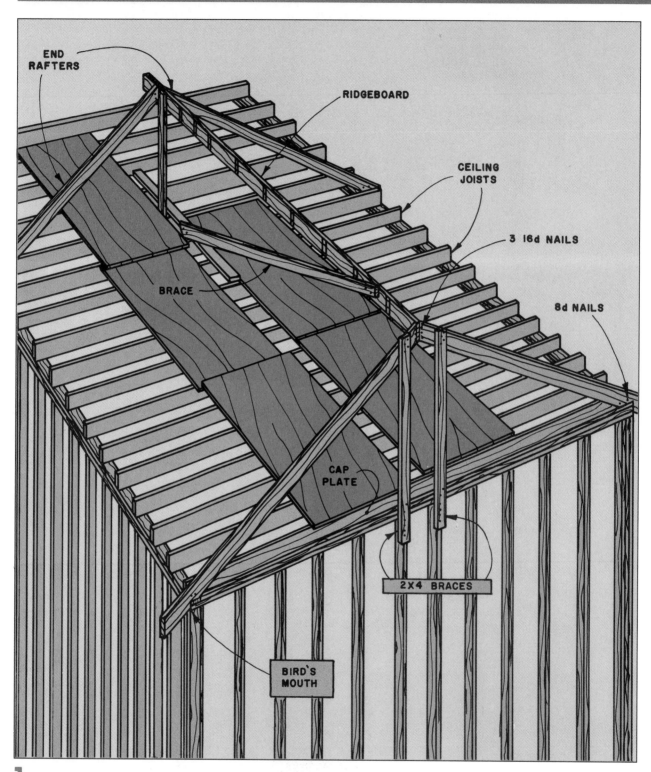

END
RAFTERS

RIDGEBOARD

CEILING
JOISTS

3 16d NAILS

8d NAILS

BRACE

CAP
PLATE

2 X 4 BRACES

BIRD'S
MOUTH

1 Use at least three people to frame a large roof. Lay plywood sheets on top of ceiling joists for support. Measure for and cut the ridgeboard, and mark it for the rafters. Nail two upright 2 x 4s flush against the cap plate for use as braces. Raise the first end rafter and align it with the bird's mouth. Toenail it to the cap plate with three 8d nails or a nail-on plate. Tack it to the top of the brace. Raise the ridgeboard into place, align it to the top of the rafter and nail it to the rafter with three 16d nails. Repeat the process at the opposite gable to raise the ridgeboard at the other end. If your roof is long, run a 2 x 4 from the ridgeboard to a piece of scrap lumber which is nailed to a ceiling joist. This brace will help to steady the ridgeboard while you attach the rafters.

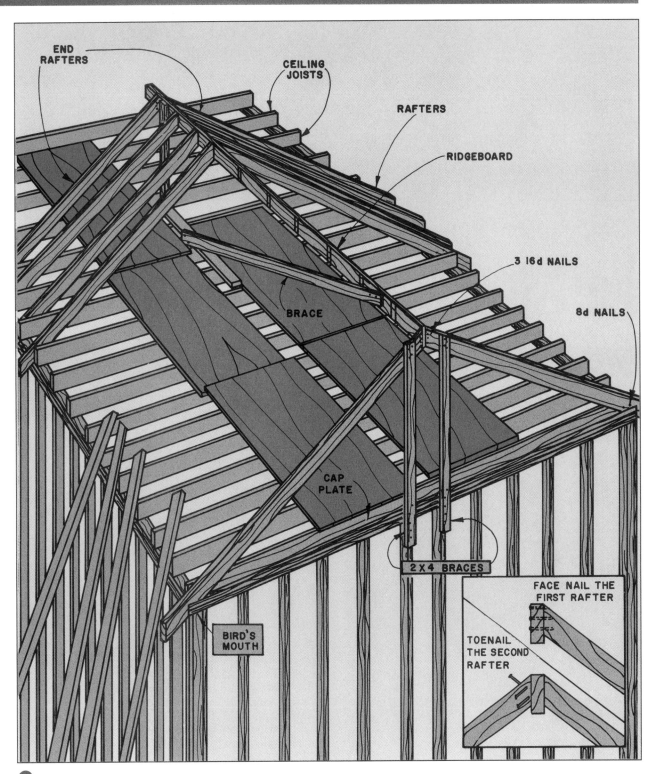

END
RAFTERS

CEILING
JOISTS

RAFTERS

RIDGEBOARD

3 16d NAILS

8d NAILS

BRACE

CAP
PLATE

BIRD'S
MOUTH

2 X 4 BRACES

FACE NAIL THE
FIRST RAFTER

TOENAIL
THE SECOND
RAFTER

2 Prop rafters against the cap plate, adjacent to their intended positions. Pull them up from below as you need them. Nail each rafter pair into place with this method: First toenail one to the cap plate, then secure it to the ridgeboard by driving three 10d nails through the ridge-board and into the rafter. (If the ridgeboard is made of 1-by lumber, use 16d nails.) Attach the rafter's mate to the ridgeboard by toenailing two 8d nails from each side.

THE FUNDAMENTALS OF FRAMING

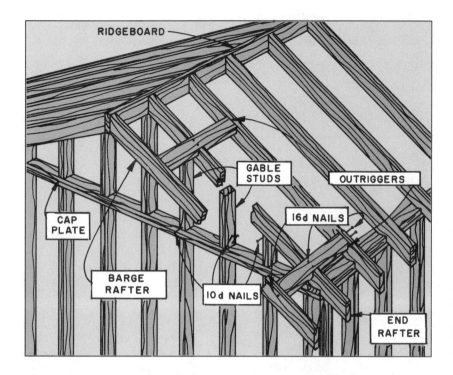

Gable End Framing. Gable studs are installed between the end rafters and the cap plate, 16 inches on center. Install the first stud between the ridgeboard and cap plate and measure in both directions for remaining studs. Toenail studs to the cap plate and rafters with 10d nails.

Use 2 x 4 outriggers to extend gable ends for an overhang. Attach them at 4-foot intervals to the next-to-the-end rafters. Notch the end rafters so the outriggers fit flush to them. Fasten the outermost *barge rafters* to the ridgeboard and the outriggers. Use 16d nails at all joints.

TWO TYPES OF SOFFITS

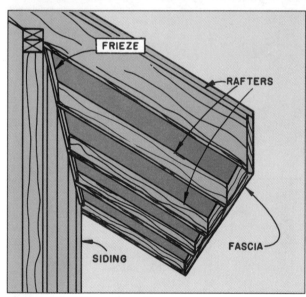

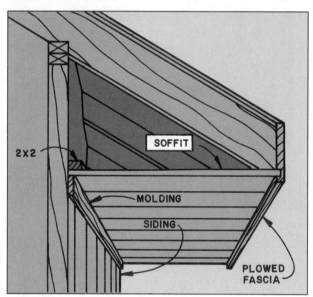

Open Soffit. The open soffit exposes the underside of the rafters and the siding to weathering, so quality materials should be used. The house siding is notched to fit around rafters; trim called a *frieze* is fastened to the top edge. Often a *fascia* is secured to the rafter ends.

Closed Soffit. Also called a *box soffit*, this type, which has many variations, is identified by the soffit board or panel which extends from the bottom of the fascia to the side of the house. The one shown here has a *plowed fascia*, a purchased board, attached to the rafter ends. It has a routed notch for holding the soffit.

To install, first put up the fascia; then find points at each end level to the bottom of the notch. Connect them with a chalk line on the wall. Above it, nail a 2 x 2; to it, fasten the soffit, and finish with a molding strip secured with galvanized nails.

THE FUNDAMENTALS OF FRAMING

1 Install a roof plate of 12-inch-wide x ½-inch plywood. Determine its position by snapping a chalk line from the point where the ridgeboard meets the roof to the top of an extra-long rafter (inset). Further in from this line, snap a second chalk line at the correct distance (see chart, page 101). Drop plumb lines from the rafter marks on the ridgeboard to the existing roof. Make marks and extend horizontal lines to the roof plate. These are the markers for the short jack rafters. Measure and cut the rafters *to approximate size* by using the same angle as was used in the horizontal plane of the bird's mouth. Cut the rafter at the correct angle (see chart page 101). Use long boards to create both the right and left jack rafters.

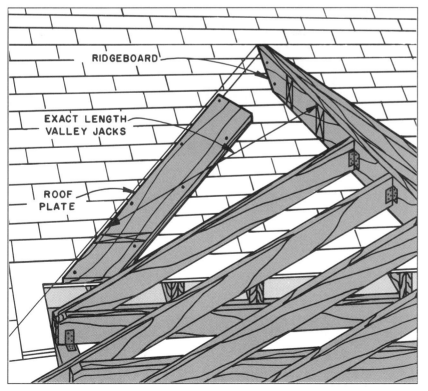

2 Measure for the exact length of the valley jacks by extending a tape measure from the top of the ridge-board to the far end of the roof plate. Mark this distance on the upper ridge of each left jack rafter. Again using the rafter template, transfer the cutline for the ridgeboard to the jack rafter (inset). Use left-side rafters as templates to cut right-side rafters. Toenail them to the roof plate and then fasten them to the ridgeboard like the common rafters (page 97).

Easy Soffits. Where the rafters meet the cap plate a cornice is created. When rafters extend beyond the cap plate a bottom plate called a *soffit* is attached. Accordingly, the term 'soffit' is used loosely to describe this area of the roof.

Soffit design, of course, should have been determined prior to the roof framing since the length of the rafters is critical. Soffits should match or complement the existing structure.

Although there are countless kinds of soffits, and some can be very tricky to build, the two very basic types are outlined here — open and closed (page 98). Beginners should work with these two relatively easy types. To make your job even simpler, purchase a fascia instead of routing one yourself. Aluminum soffits are also available.

Adding the Sheathing. When the roof framing is completed, you should cover it with a smooth covering called the *sheathing*. This can be of several materials; plywood, common boards, shiplap, and nonwood are all used. The sheathing provides a nailing surface for the roofing that's to come. Depending on geographical location and your choice of roof covering, the sheathing can be installed solidly across the roof (with no gaps) or spaced apart in strips.

Your local building code will give you guidelines about what type of sheathing to choose. A very popular type is plywood. It makes a firm base for most types of roofing and its availability in large sheets makes for fast application. When you install this type, lay it out just like the subflooring — with the length of the sheets running across rafters.

Choosing a Shed or Gable Roof. When you add onto an existing structure, the easiest type of roof to use is the shed roof. Attach it directly to the exterior wall of the existing structure and simply add the rafters. Use the instructions below and on page 101 to frame this type of roof. An option is to delete

ADDING ON WITH A SHED ROOF

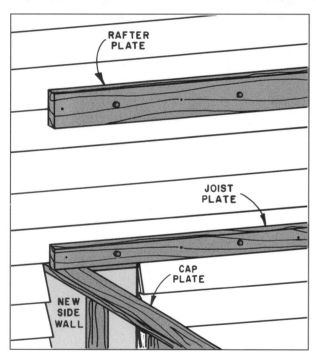

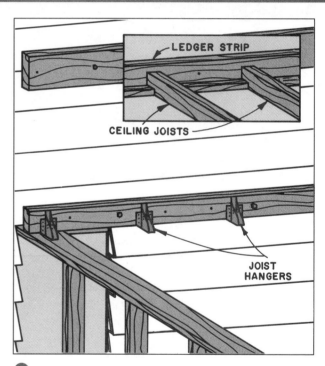

1 Mount two plates to the existing structure. Cut the joist plate the same size as other joists; it should span from the end of one cap plate to the other. The rafter plate should be as wide as the cut end of the rafters. Determine where to place it by subtracting 5½ inches from the length of a side wall, dividing by 12 and multiplying by the roof slope (example: use 6 if the roof is 6 in 12). To this figure, add the dimension of the vertical end cut. Fasten plates with nails and then drive lag bolts into the wall studs.

2 Secure metal joist hangers to the joist plate so their bottom edges are flush to the bottom of the plate. Position them 1½ inches from the ends of the plate and at 24-inch intervals. An option is to use wooden ledger strips and toenailing to support the joists (inset). Using corresponding measurements, secure metal joist anchors to the cap plate at the other end of the structure. Measure, cut, and install ceiling joists.

THE FUNDAMENTALS OF FRAMING

the ceiling joists and create a slanted ceiling.

A gable roof also can be attached to the side of an existing structure. When a new gable roof meets the slope of an existing roof, the process is a little tricky. It involves installing a plate and attaching short rafters called *valley jacks* to the existing roof. Here you'll have to use your measuring and marking skills even more.

In order to make the valley jacks fit snugly, you'll have to know two things: exactly where to position the roof plate and the precise angles at which to cut the short rafters. Using the principles of roof slope (pages 68-71), consult the proper line in the chart to the right. It will tell you how far back to nail the plate and the angle at which to set your saw when creating the jack rafters. Follow the instructions on page 99 to add a gable roof to an existing structure.

FITTING JACK RAFTERS TO THE ROOF

Slope (1 in 12, etc.)	Setback of Plate	Rafter Angle at Foot
1"	4¼"	5°
2"	2⅛"	10°
3"	1⅜"	15°
4"	1"	19°
5"	⅞"	23°
6"	¾"	27°
7"	⅝"	31°
8"	½"	34°
9"	½"	37°
10"	½"	40°
11"	⅜"	43°

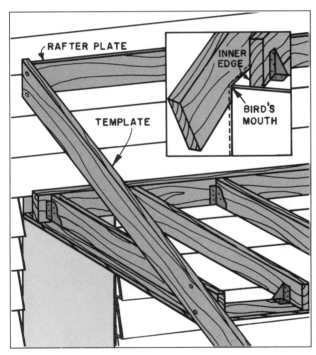

3 Make a template for all rafters with a long board. Cut it at a rough angle to match the roof slope. Temporarily nail it to the rafter plate with its upper edge level to the upper edge of the plate. Then nail the bottom of the 'rafter' as follows: Align its lower edge with the inner edge of the cap plate. Mark it for the bird's mouth by tracing the outside wall. Trace the other rafter end for an accurate plumb cut and add the mark for the tail cut. Cut the template board and use it to cut rafters.

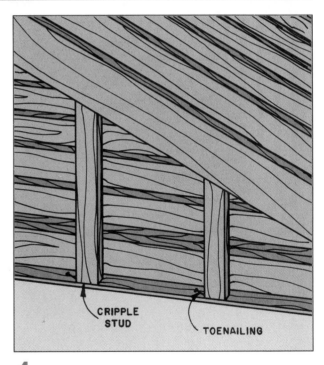

4 Fasten the end rafters; set them flush to the end ceiling joists. Nail with metal reinforcing plates. Install inside rafters to match joists, spaced 24 inches apart. Cut 2 x 4 cripple studs to fit between the side wall cap plates and the outer two rafters. Align them 16 inches apart with faces parallel to the joists; toenail at top and bottom. Nail studs to the exterior wall — between the rafter plate and joist plate with faces flush to the wall. Finish by cutting the ceiling joist ends flush to the upper edges of the rafters.

7

Roofing and Siding

As soon as the walls and roof of your structure are framed, you'll want to waste no time getting the framework covered. Adding the roofing and siding can be fun since you'll be venturing into the realm of exterior decorating — choosing the textures and colors that express your individuality as well as match or coordinate with your existing structure. It's wise to do the roofing first to protect the inside framing and to provide a covered workspace if it rains.

Roofing materials must be selected and applied with care. In fact, the type of final surface that you choose requires a particular kind of base in the form of sheathing and underlayment materials. Drip edges are installed at the edges for protection from water, and flashing must be installed to route streams of rooftop water. Since all these materials overlap each other in a specific way, you should have a plan and a clear understanding of what each step entails before you begin a roofing project.

There are a multitude of exterior siding materials to choose from. But here you'll learn about only the most popular and easiest types to install. Depending on where you live, you'll want to choose siding that gives you extra protection from the weather and helps to keep your energy bills at a reasonable level. To protect the siding, you'll need to use the proper finish. Exterior finishes are described and you'll learn how to apply them.

As part of this chapter, you'll learn how to easily install prehung doors and windows. A helper, some shims, and an accurate level are all that's needed.

Finally, you'll see how to work with gutters and downspouts. These are put up after the roofing, siding, and finishing is completed. When you're all done, the house will be 'dressed' for the weather; then *you* can move indoors.

Roofing — Covering the Rafters

Basically, there are four steps to installing the roofing:

1. Lay the sheathing.
2. Apply the underlayment.
3. Install the drip edges and flashing.
4. Lay the roofing material.

These will be covered in chronological order, although some of the processes will 'overlap'. The following are the basic elements for roofs with fiberglass and asphalt shingles, and wood shakes (below).

Sheathing. When the roof framing is completed, you should surface it with a smooth covering called the sheathing. This is usually plywood or common boards. The sheathing provides a nailing surface for the roofing materials that are to come.

Depending on your choice of exterior roof covering, the sheathing can be installed solidly across the roof (with no gaps) or spaced apart in strips. Shingles and wood shakes are laid over solid sheathing, whereas wood shingles are commonly installed over spaced 1 x 4 boards. Your local building code will give you guidelines about what type of sheathing to use.

The most popular type of sheathing is plywood; its availability in large sheets makes for fast application. When you install this type, lay it out just like the subflooring — with the length of the sheets running across rafters. Often, the shape of the roof will not allow for perfect rows of rectangular sheets, so be sure to measure and cut them as needed.

Use ½- or ⅝-inch plywood for a strong nailing base. Leave a ⅛-inch gap between panel ends and nail all ends centered on rafters. Leave a 1/16-inch gap at panel edges; double both of these gap dimensions if your weather is unusually humid.

ROOFING COMPONENTS

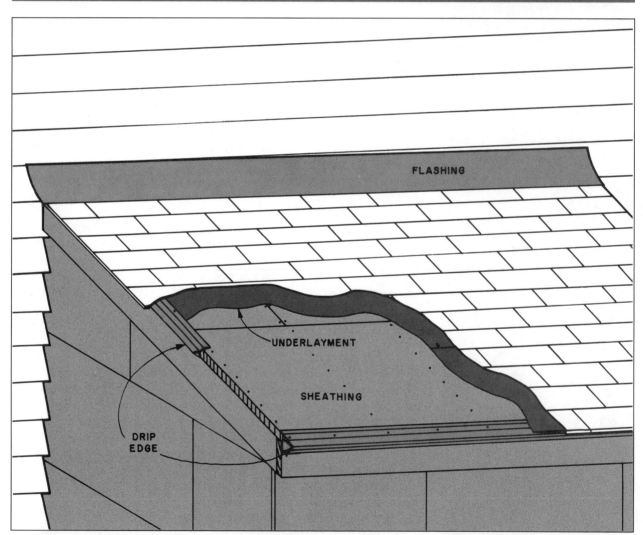

These basic components are used when the roof is finished with asphalt shingles or wood shakes. The *sheathing* is made of plywood or particle-board sheets nailed horizontally across the rafters. On top of it is nailed the *underlayment*, a felt material that comes in rolls. *Flashing and drip edges* protect valleys, edges, and other places from water seepage.

1 The individual sheets of sheathing are handed up through the rafters from below. The person on top pulls the sheet out.

2 Note how the sheathing is nailed horizontally, and staggered across the rafter. Use special clips (inset) to keep the sheets aligned. This makes for an even more 'perfect' roof when settling occurs.

3 On top of the sheathing, felt underlayment is nailed.

ROOFING AND SIDING

Do the fastening with 6d nails for ½-inch thick plywood and 8d nails for ⅝-inch thick plywood. Position the nails every 6 inches along vertical ends of panels and 12 inches apart at intermediate supporting rafters. For a good smooth roof, use special clips or fasteners that keep the sheathing panels aligned.

Underlayment. Next, roll out the underlayment — a felt-like material that's purchased in rolls. Underlayment serves several purposes. It protects the sheathing from rain before the roofing material is installed and from wetness if the roofing material is raised during a storm. The weight of the felt is contingent upon the material that will be applied on top of it. If you're going to use shingles, use 15-pound felt. For wood shakes, use 30-pound felt.

Use the following method to install underlayment for a roof that has a slope of 4 or more inches in 12. (A flatter roof also requires underlayment but the installation method will be different.) As always, check your local building code for the specifications in your area.

Roll out the felt horizontally working from eave to ridge. Begin by snapping chalk lines. The first chalk line should be at 33¼ inches to allow for the ½-inch overhang over the eave. Successive chalk lines should be positioned at 34 inches to allow for the 2-inch overlaps at the start of each new row. When the roll runs out, overlap sides by 4 inches. At the overhang, trim the felt flush to the edge. Note that here the felt will go *under* the drip edge, but at eaves, it will go *over* the drip edge. At any hips, ridges, or valleys, overlap the felt by 6 inches.

Use 1¼-inch galvanized roofing nails for the fastening; drive only enough nails to hold the

INSTALLING ROOF FLASHINGS

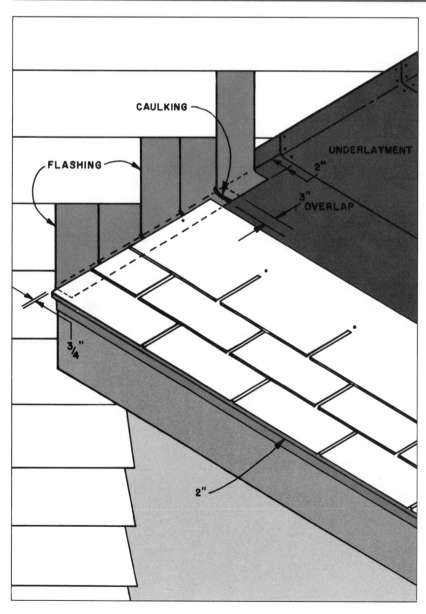

Where a Gable Meets a Wall.
Apply flashing alternately with courses of roofing. Cut small rectangular pieces of flashing and bend them to fit on the roof so at least 2 inches will be laid under courses of roofing. Fit pieces underneath wall panels, using 3-inch overlaps as shown. Caulk vertical gaps between flashing. At the roof peak, use two pieces of flashing. Fold one over the ridge and extend it up the wall on each side. Extend the second piece down the wall over the first and notch it to fit the roof peak.

underlayment in place until the roofing is installed (approximately one nail per square yard of felt).

Drip Edges and Flashings. After nailing the felt, the next step in covering the roof is to install flashings, special coverings, over potential trouble spots. Some types of flashings, however, should be installed *while* the roofing is laid. Accordingly, determine which areas of your roof will need flashing before you start laying shingles or shakes.

The most common type of flashing material is 28-gauge galvanized sheet metal. Other metals include aluminum and copper. Plastic flashings are available, and mineral-surface roll roofing can be used in certain places on shingle roofs. Preformed flashings are widely available, or you can shape your own.

Flashings are required at any point in the roof where joints can cause water leakage and potential damage to the sheathing. Your local building code will detail where this extra protection is needed. Basically, edges, vents, chimneys, valleys, and joints at vertical walls are the most vulnerable spots.

Drip edges are required at overhangs and eaves. As mentioned above, make sure that felt is installed properly in relation to the drip edges. Vent pipes and chimney flashings take special care to install; they are put on at the same time as the roofing or woven with it. Caulking or plastic cement may be used with these as well as with other flashings.

Two ways to install flashings are shown on page 106 and this page. One outlines the process to use where a gable meets a wall. The other shows how to install valley flashing where a gable intersects an existing roof.

Roofing Materials. Now it's time to give the roof its exterior coat of roofing. Of course, there are

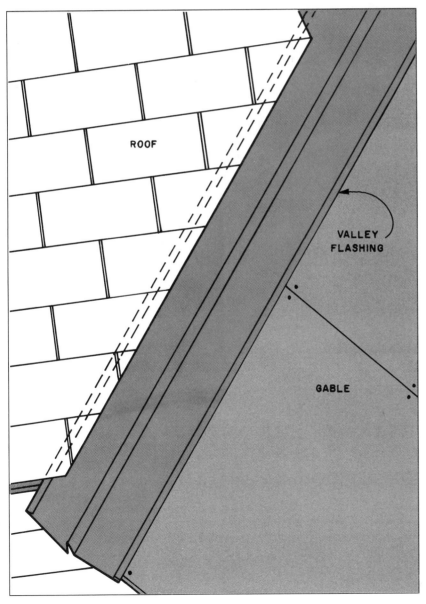

Where a Gable Meets an Existing Roof. Use a special type of valley flashing which includes cleats for nailing. Cut back existing roofing 3 inches from the center of the valley and remove any nails 7 inches from the center. Remove roofing material at the peak to expose about 2 square feet of sheathing. Slide the valley flashing under the shingles or roofing material and nail it to the roof through its cleats. At a distance 1 foot above the peak, trim the flashing and nail it flat to the roof. Replace the shingles or roofing material.

many options in choosing materials, and there are various installation methods as well. But for your purposes as a beginner, only the most common and the easiest materials and methods will be covered.

Keep in mind that the most attractive look will be achieved by matching the existing roof. This can be quite difficult if the existing roof is aging because it might be faded. If the existing material is very old, you might consider roofing the entire structure.

Two popular forms of roofing are covered — shingles and wood shakes.

Fiberglass and Asphalt Shingles. Perhaps the most popular type of roofing, shingles come in a wide array of colors and shapes. The easiest type to work with are the rectangular sheets that have two cuts in

them to create three tabs. These commonly measure 12 inches by 36 inches. Each shingle has a strip of self-sealing mastic just above the tabs that adheres to the shingle tabs laid on top of it.

Shingles must be laid with the correct *weather exposure* — that is, the correct amount of surface exposed to the elements. Typically, the weather exposure for asphalt shingles is 5 inches.

Determine the appropriate size fasteners according to what the shingles will be installed over. Generally, use 12-gauge galvanized roofing nails with ⅜-inch-diameter heads for the job.

Before you begin, sweep the underlayment clean and check it for any protruding nailheads. Then begin to install the shingles as shown below. Work

INSTALLING SHINGLES

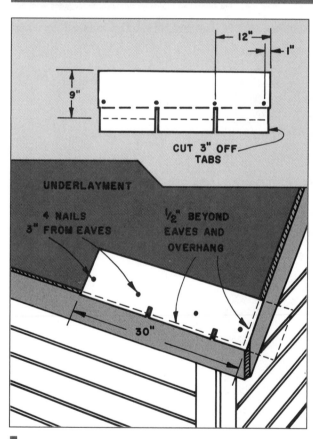

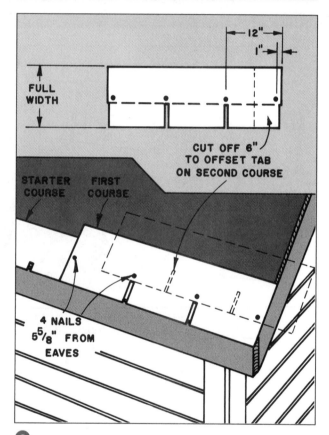

1 Run the *starter course* from over-hang to overhang with 9-inch wide material. Either use roll roofing or create your own starter course by cutting 3 inches off the tabs of 12-inch shingles. Starting from the left, lay the starter course with self-sealing strips down. Allow ½-inch to extend beyond the eaves and overhangs, and ¹⁄₁₆-inch gaps between shingles. Position the nails 3 inches above the eave; drive two nails 1 inch from each end and two nails 12 inches from each end.

2 Run the *first course* with full-width (uncut) shingles. Lay them directly over the starter course so they also extend ½-inch beyond the eaves and overhangs. Again place the shingles ¹⁄₁₆ inch apart. Position nails 5⅝ inches above the eave, with two nails 1 inch from the ends and two nails 12 inches from ends — or as specified by the manufacturer. In order to offset the tabs of the next course, cut 6 inches off the end of the first shingle.

from one end to the other unless the roof extends further than 30 feet; in that case, start in the center. Use chalk lines faithfully and work in *courses* or built-up rows from bottom to top. Use the guide below when you reach the hips and ridges.

Wooden Shakes. If you've got plenty of time or you prefer a rustic look, you'll want to install wooden shakes. These individual 'slats' of wood should not be confused with wooden shingles which are thinner and call for a different application method.

Shakes are split by machine or by hand from chunks of western red cedar. Some shakes are tapered at one end, while others are equally thick throughout. The thin end, or in the latter case, the smooth end is always laid uproof. Shakes, which can also be used for exterior wall siding, come in 18- or 24-inch lengths. Many grades are available, but for roofing 'Blue Label' or Number 1 is recommended.

Rustproof nails are necessary for laying shakes. Choose 13-gauge nails with 7/32-inch-wide heads. As for length, make sure that they will penetrate 3/4 inch; 2-inch nails should suffice.

The exposure for shakes is determined by the pitch of the roof and by the length of the shakes. If your roof is 4 in 12 or steeper, use an exposure of 7½ inches for 18-inch shakes, or an exposure of 10 inches for 24-inch shakes.

Install shakes according to the instructions on page 110. As previously mentioned, the roof sheathing may be of spaced boards instead of the solid plywood shown. Chalk lines are used to align shakes horizontally but none are required for vertical alignment. Instead, offset the shakes 1½ inches to ensure that no three rows have matching joints. Since shakes are made of wood, slight gaps are required between them to allow for expansion and contraction. For extra protection, 30-pound felt underlayment is laid under each course.

When you need to trim shakes use these methods: For a straight cut, simply use a roofer's hatchet. For an angled cut at a valley, lay the shake in place and mark the angle with a straightedge. Then, use a saw to make the cut.

Covering the Studs — Exterior Siding

Equal in importance to covering the roof is the process of installing a covering over the wall studs. Again, there are countless choices — both of wall siding and of sheathing. Depending on what your final siding will be, you should select a coordinating sheathing or nailing surface. With this in mind, use these basic steps to do the wall siding:

1. Install the sheathing.
2. Add the building paper (if necessary).
3. Install the windows and doors.
4. Install the siding.

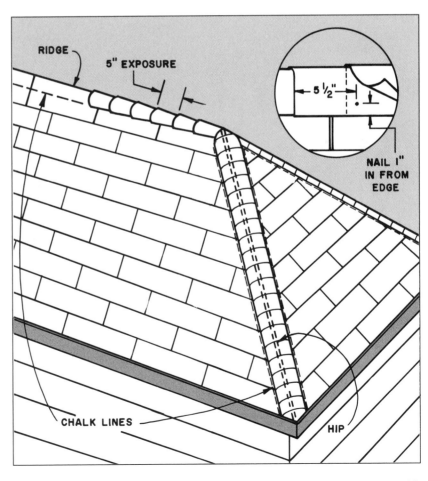

Installing Shingles at Ridges and Hips. Snap chalk lines 6 inches on center at both sides of the hip(s) and ridge. Using either special shingles or hand-cut 12-inch-square pieces, begin nailing at the hip. Apply a double layer at the bottom and work upward, aligning shingles with the chalk lines and leaving a 5-inch exposure. If you're using handmade pieces, bend them as needed to fit snugly. In cold weather, warm the shingles first. Nail each shingle with one nail — at 1 inch from the end and 5½ inches from the butt.

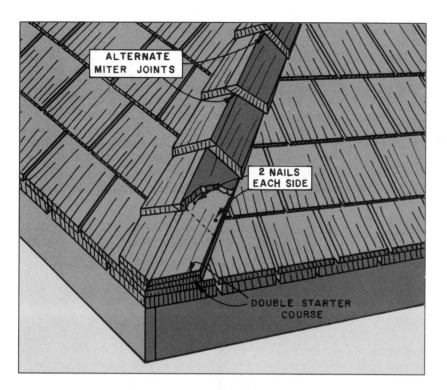

Installing Wood Shakes at Ridges and Hips. Use factory-made mitered shakes. Install the row opposite prevailing winds first using the same exposure as was used on the roof planes. Double the starter course at ridge ends and hip bottoms. Alternate the miter joints, as shown. Use nails long enough to penetrate the ridgeboard — either 2 or 2½ inches. Drive two nails into each shake, positioning them as shown.

INSTALLING WOOD SHAKES

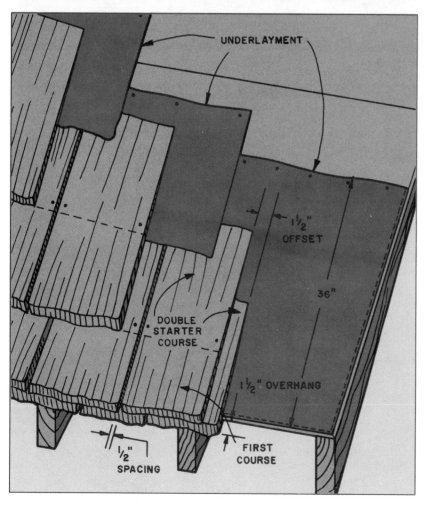

Begin by laying a 36-inch-wide strip of underlayment along the eave; make a ⅜-inch overhang. Install a double starter course of shakes with the ends extended beyond the eaves and overhangs 1½ inches, and with the top shakes offset 1½ inches. Nail the shakes with 2 nails — both placed 1 inch from the butt line and 1 inch from the edge. Leave ½-inch gaps between shakes and use underlayment as shown. First, measure for the underlayment. From the butt of the double starter course, snap a chalk line that measures twice the distance of the exposure. Lay the bottom edge of 18-inch-wide underlayment at this line, then nail along the top edge every 12 inches. When the roll runs out, overlap the underlayment 4 inches. Next, snap a chalk line on the first course to indicate exposure. Nail the next course, lay the next underlayment, and continue until you reach the ridge. Lay 15-inch shakes for the last courses; trim them all at once and then lay an 8-inch-wide strip of underlayment.

ROOFING AND SIDING

All About Wall Sheathing

Just as the roof required a sheathing, the wall studs usually need a similar 'protective layer'. An exception is walls which are covered with very thick siding — usually a vertical type. In addition to serving as a nailing surface, the sheathing ideally adds extra bracing. Some types of sheathing even have insulating properties.

Products and brand names vary but these are the four 'generic' types of sheathing: solid lumber, plywood, gypsum board, and fiberboard. Though nailing methods vary, generally the use of rustproof nails is recommended and the nails should penetrate the surface at least 1 inch. Choose among sheathings according to the final covering, ease of installation,

and amount of insulation required. Check your local building code for recommendations.

Solid Lumber. This type of sheathing, usually 1-inch nominal boards measuring from 6 to 12 inches wide, obviously will take a while to install. The process is speeded up slightly by shiplap or tongue-and-groove patterns — on the board sides and sometimes on the edges. Boards may be installed horizontally or diagonally. With the horizontal method, stud bracing is required.

Boards are nailed to studs and are patterned so that the ends meet studs on center, unless the aforementioned end-matched kind are used. The number of fasteners is determined by the width of the boards. Use two 8d nails driven into each stud for 6- or 8-inch boards or three of the same for 10- or 12-inch boards.

TYPES OF WALL SHEATHING

Plywood sheathing. Plywood is the most popular sheathing material. Although it may be installed horizontally, the most preferred method is vertical. If you install panels horizontally, add blocking between the horizontal joists as bases for nailing.

Three other kinds of sheathing. The foil-covered type has the highest 'R-value' which means that it has the best insulation properties, so it is used to conserve energy at the main part of this house. The black type, on this structure, is used for the garage where there will be no heat. Particleboard or plywood is used at the corners as a wind brace, and also to keep the corners square.

Plywood. Because plywood is strong and comes in large 4 x 8 foot sheets, it's a very popular sheathing material. When you purchase it, be sure to specify 'exterior grade' or 'performance-rated'. Choose regular square-edged panels, or, as an option, easy-to-use panels with tongue-and-groove edges.

Although plywood comes in several thicknesses, the ones most often chosen are 5/16 inch and 3/8 inch. If the exterior finish is to be nailed to it, use the thicker 3/8-inch sheet. Plywood may be nailed either horizontally or vertically, but the vertical method is much preferred for rigidity.

Use 6d nails positioned every 6 inches at all edges of the sheets; position the nails 12 inches apart when nailing into studs. Slight gaps are recommended between panels — at horizontal edges 1/8 inch apart, and at ends 1/16 inch apart.

Gypsum Board. This type of sheathing, under various trademarks, is actually like an insulated subsiding. It is composed of fiberglass-impregnated gypsum sandwiched between water-repellent, usually foil-like, paper. Gypsum board is the best of the sheathings when it comes to insulation value. One of the drawbacks in using gypsum board is that you will probably need to brace wall studs diagonally for stiffness. Another is that as a nailing base for siding, it is unsuitable; so you'll have to choose a type of siding that can be attached to studs with long nails. Widely available in 1/2-inch thicknesses, the usual type is 2 x 8 foot tongue-and-groove panels.

For this material use special 'gypsum' nails. Some have wide plastic 'caps' at their heads to provide extra holding power. Place them 4 inches apart at all edges and 8 inches apart at intermediate studs.

Fiberboard. Fiberboard, like gypsum board, is not a strong material, so its use also dictates additional stud bracing. It can be used as a nailing base, however, as long as the proper grade is specified.

This sheathing is readily identified by its black surface. Though products vary, it is generally manufactured of wood pulp, cornstalks, sugar cane, and the like. This is pressed into 'boards' either 25/32 inch or 1/2 inch thick. These are impregnated with asphalt for water resistance. Although sizes vary, homeowners usually choose 2 by 8 foot panels.

When nailing fiberboard, use special nails or roofing nails and space them 3 inches apart at all edges and 6 inches apart at intermediate studs. Choose 1 1/2-inch nails for 1/2-inch boards or 1 3/4-inch nails for the thicker boards.

Building Paper — An Extra Coat

The next step, after putting on the sheathing, is to apply what is commonly called *building paper* or *sheathing paper*. (This step might *not* be needed, however, if you used gypsum board or fiberboard — both good insulators — for the sheathing.) Building paper is strongly recommended for sidings that have many

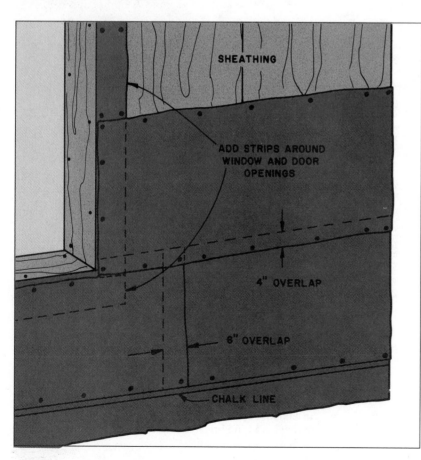

SHEATHING

ADD STRIPS AROUND WINDOW AND DOOR OPENINGS

4" OVERLAP

6" OVERLAP

CHALK LINE

Using Building Paper. Building paper, required by many codes, is used with some types of sheathing or it may be used alone between studs and siding. Install it horizontally from bottom to top. Snap a chalk line for the bottom edge of the first course. Overlap paper 6 inches at vertical joints and corners, and 4 inches at horizontal joints. Staple or nail it to the surface just enough to hold it in place until exterior siding is installed. Add strips of building paper to window and door openings.

joints such as narrow boards, shingles, or shakes. It takes on even more importance in areas subjected to storms and heavy winds. Local building codes will guide you in the use of building paper.

The purpose of building paper is to *resist* water. It should not be strictly waterproof since the creation of a vapor barrier will cause condensation problems. Materials differ but it is generally manufactured of felt or kraft paper which is impregnated with asphalt.

Install building paper as soon as possible after the sheathing is up. An easy process (page 112), all that's required is that it be tacked up. When the final siding is applied, it will be simultaneously fastened to the wall.

Windows and Doors — Working with Prehung Units

Before the siding is installed, you'll need to fill in the cavities of the framework with windows and doors. Traditionally, these procedures were long and tedious, plus they took considerable experience to accomplish. Not so today; prehung windows and doors are the norm and they're relatively simple to install. Before you begin, however, make sure that you have followed the framing instructions for the rough openings (pages 86 and 87).

Factory-Made Windows. Complete windows of all types are delivered ready to install; so you need not feel limited in your choice. The only thing that probably won't be included with a prehung window is the interior casing. The three basic types of windows are fixed, sliding, and swinging. Although there are many variations within each type, the most popular window is the vertical sliding kind called a *double-hung* window. Its components are shown below.

Double-hung windows are available in different styles according to how they are to be mounted. Some are made to be nailed through the exterior trim of the window while others are nailed through a flange that surrounds it. In other styles the jambs are nailed or screwed through shims into the trimmer studs and the header.

Always check the manufacturer's instructions to determine the best nailing method for your window. Often, they will recommend a preliminary step of lining the cavity with building paper or plastic sheeting. It should be noted that there are many, many ways to install windows. They may be screwed or nailed into place. Installation may be from the interior or exterior of the house — or both. Basic prehung window installation is shown on pages 114 and 115.

Generally, installation involves tacking the window while checking it for level and plumb and shimming it. Once this is done, the primary nailing should be done with finish nails.

Ready-to-Hang Doors. The time-honored method of installing a door involves these steps: constructing a door frame, attaching it on the wall

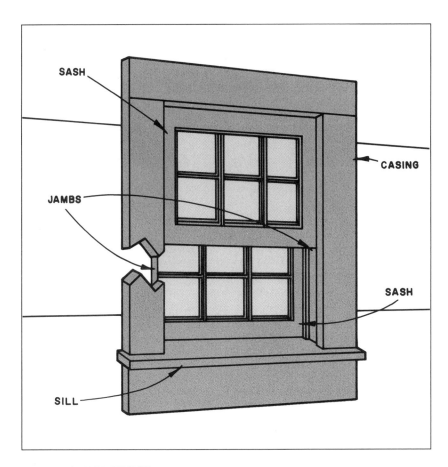

Window Components. Use this illustration of a *double-hung* window, the most popular type, to identify window components. It includes two movable *sashes* — an upper one on the outside and a lower one on the inside.

1 Always read the manufacturer's instructions as window products vary greatly. Install the windows after the sheathing is installed and before the exterior siding is added.

3 Drive in shims to correct the position of the window.

4 Install the screws. These windows have caps that fit over the screws. If you're working with windows that don't, then use 'attractive' screws such as Phillips head or white ones. Four screws were used here — two at the top and two at the bottom. Put them in the snug side first if one side is looser than the other. After all screws are installed, remove the packing strap.

2 Check the window for plumb to see how it fits into the opening.

5 An important part of any window installation is the caulking process. This provides extra protection against the weather.

inside the rough opening, hinging the door to the frame, and installing a lockset or latch. Although this process yields a truly customized passageway, the novice can get similar results much more quickly by using prehung doors.

Doors are available in two basic types: *panel* and *flush*. The panel type is normally used as an exterior door since it is sturdy and attractive — named for its panels that are set between rails and stiles (page 116). The flush type is made from flat face and back veneers, usually of 1/8-inch plywood. These are attached to either a solid core or a hollow grid-like core. The hollow-type door is strictly for interior use.

Similar to window installation, the process of putting in a door requires shimming and consistent checking of level and plumb. There are, obviously, two kinds of door installations — interior and exterior. Products will vary so be sure to follow the manufacturer's recommendations. Basically, installation consists of several steps (page 117).

Siding Simplified

After the windows and doors are installed, it's time to fasten the siding to the exterior walls. Choose your wall siding by the following criteria. Match or coordinate it to the existing structure. If you're a beginner, choose a siding that can be installed easily and quickly. If you want low-maintenance in the future, select a material that's prefinished.

Popular choices include solid boards, vinyl siding, aluminum siding, exterior plywood, and hardboard. The first three are typically installed horizontally and the last two are installed vertically.

Solid Boards. These are the 'classic' wood sidings that lend a truly natural look to your handiwork. Boards are available in a large variety of materials. Sizes commonly vary from 4 inches to 12 inches in width. The drawback, of course, is maintenance. If you choose wood board siding, avoid termite damage and water rot by using redwood or cedar.

Sheathing must be used with board siding, either in the form of plywood, gypsum board, or fiberboard, or with a combination of lumber and building paper. It's best to check your local building code for the best combination. Although there are many variations in installation methods, the two basic kinds are *rabbeted bevel* and *shiplap*. Both of these methods are illustrated on page 118.

Vinyl and Aluminum. These low-maintenance products rate very close to each other in durability, price, and ease of installation. Vinyl siding is a bit easier to install than aluminum and it won't show dents or scratches. However, it does become brittle in cold weather and it will fade in time. Aluminum may corrode near salt water but unlike vinyl siding it can be painted. Both products require a base of sheathing as well as building paper. They can be installed with the help of a few specialty tools, lots of time, and a complete understanding of the manufacturer's directions.

Exterior Plywood and Hardboard. The chief advantage of using these vertical sheet materials is that they can be installed very quickly. Large panels measure 4 feet wide by 8 to 10 feet long. Exterior plywood needs to be finished; a popular type has grooves in it to simulate solid board siding.

Hardboard comes in a variety of surfaces which range from rough wood to prefinished stucco. With some of these vertically-installed materials, sheathing is not required. Consult your local building code to find out if unsheathed walls will give you adequate insulation. Depending on the siding, various joining methods are used, but regardless, the fastening should always take place at wall studs. Additional caulking is usually recommended.

Adding Protection – Finishing the Exterior

There is a wide variety of exterior wood finishes to choose from. In fact, the sheer number of choices may at first be confusing. However, making your selection is easy once you understand a few important principles about outdoor finishes.

How Exterior Finishes Are Made. Most finishes are a mixture of two substances — *pigments* and *vehicles*.

Pigments give paint or stain its color. The pigment chemicals change from finish to finish, depending on its color or lack of it. When choosing a finish, the pigments deserve very little consideration. You only need to buy a reputable brand and make sure that the colors won't fade.

The vehicles in the finish are much more important. These bond the pigment to the wood — and keep it there through all sorts of weather. Vehicles are typically resins, or emulsions. They have many characteristics, such as penetration, coverage, and drying time. The most important characteristic is the *hardness*.

Hardness and Hardness Rating. Generally, the harder the vehicle dries, the longer the finish will last, and the better it will wear. Softer vehicles are easily abraded by wind and rain, making the finish thin. A soft vehicle, however, has its advantages. It easily contracts and expands with wood during changes in temperature and humidity. Hard-vehicle finishes tend to be brittle; they may chip or peel when the wood breathes. A good rule is to choose a soft

DOOR COMPONENTS

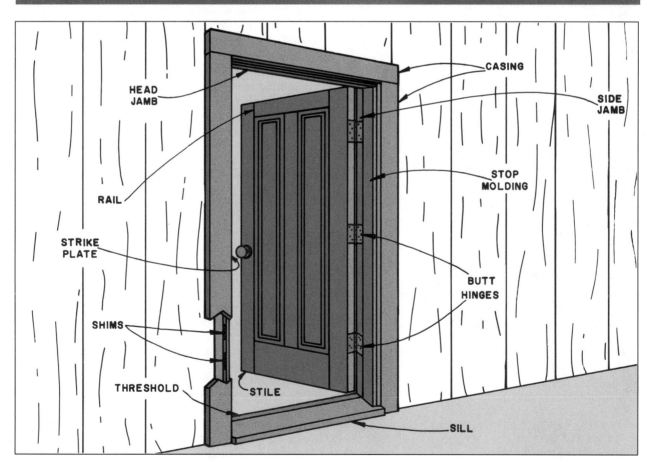

Use this illustration of a *panel door* to identify common door components.

1 In preparation for hanging the door, cut with a handsaw and remove the sole plate from the door opening. Prehung doors vary, so always read the manufacturer's instructions. The interior door shown here is one of the easiest types to install. Begin by removing any blocking, bracing, packing, or extraneous members tacked to the unit.

2 Mount the door inside the rough opening.

3 Use wedged shims to fit the door snugly. Insert the shims in pairs from opposing sides. Place them under the side jambs to meet the finished flooring, at the head jamb, at the hinge locations, and also at the strike plate location.

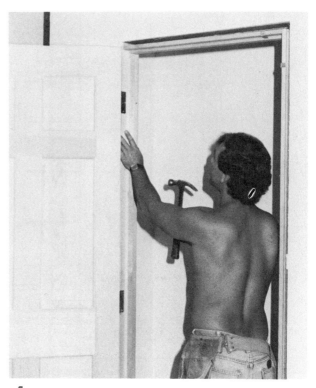

4 Check for level and plumb. Also check to see that the narrow space between the door and the jamb is equal all around. With 16d finishing nails placed approximately 16 inches apart, fasten the prehung unit at intervals and at shims; sink the nails with a nail set.

vehicle if you live in an area where the difference in average winter and summer temperatures totals more than 30° F. If the temperature changes are less radical, you can use finishes with harder vehicles.

The following three basic types of finishes — paints, stains, and clear finishes — are rated according to their hardness from Very Hard to Soft. Additional comments are given for each particular type. After reading this section, you should have a basic understanding of what type of exterior finish to use. Once you have a type in mind, you can select a brand that seems suitable. Most paint and hardware stores have displays and charts to guide you even further in your search for the ideal finish. Note: The finishes described below are for wood. If you need to apply a finish to metal, masonry, or another type of material, you might need to use a different type of finish.

Paints. Exterior paints have enough pigments mixed in the vehicle to make the mixture appear opaque. After the paint dries, all you see is the color, not the wood.

■ **Linseed Oil Paint.** Soft. Despite the soft vehicle, many experienced painters consider this to

be the best exterior paint for the price, especially in northern climates.

■ **Alkyd Resin Paint.** Medium to Hard. Alkyd is a soybean product that can be used in practically any environment — except one where there are extreme differences in temperature and humidity.

■ **Latex Emulsion Paint.** Medium to Hard. This type of paint has become popular because it's very durable and easy to clean up with water. Although it's a good general-purpose paint, it tends to peel easier than other types in a wet environment.

■ **Epoxy Resin Paint.** Very Hard. Epoxy resin paint is most useful in industrial environments where a coating that is very resistant to chemicals is needed. This paint is too hard to be used on wood in humid climates.

Stains. Stains are mixed just like paints, except they have less pigment and more vehicle. They have enough pigment to color the finish, but not enough to make it opaque. After the stain dries, you will still see the wood grain through the vehicle.

■ **Linseed Oil Stain.** See "Linseed Oil Paint."

■ **Alkyd Resin Stain.** See "Alkyd Resin Paint."

■ **Latex Emulsion Stain.** See "Latex Emulsion Paint."

■ **Varnish Stain.** Medium. Varnish is a blend of resins from several trees: gum, chinawood, and tung. A varnish stain can be used in most environments, but might not dry well in high humidity.

■ **Polyurethane Stain.** Hard. This is a synthetic resin made from petroleum. It's very durable, but unfortunately is too hard to use in environments when there are wide swings in temperature and humidity. Another drawback is that it may decay in sunlight, unless there is an ultraviolet light blocker or an absorber added.

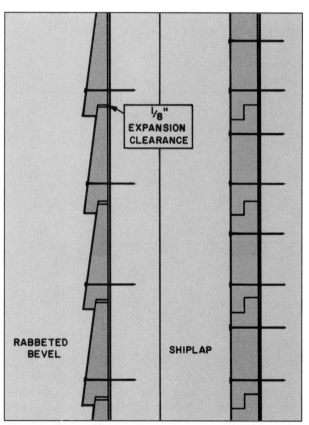

Installing Wood Siding. Note the nailing patterns for these two basic types of siding. With the rabbeted bevel type the nails are placed to just clear the edge of each preceding course. A ⅛-inch clearance allows for expansion due to weathering. Shiplap siding requires two nails, one each at the top and bottom of each course. Non-corrosive nails are required. To install ½-inch siding over plywood or board sheathing, use size 6d nails, or 8d over fiberboard. To install ¾-inch siding, use 7d nails over plywood and board sheathing, or 9d over fiberboard.

□ **WOODWORKER'S TIP** □

Because stains are simply thin paints, some professional painters save money by mixing their own stains from paints. If you wish to do the same, a standard recipe is one part paint, one part thinner, and one part "conditioner." Use linseed oil as your thinner for oil paints, and water for latex paints. Conditioners for both latex and oil paints are available at most paint stores.

Clear Finishes. Clear finishes are vehicles with no pigments. They may darken the wood slightly, but they do not change the natural color. There are two basic types of clear finishes: *coating* and *penetrating*. The former clings to the outside of the wood, covering the surface. The latter soaks into the wood, bonds with the cells, and protects it from the inside out. There is no hardness rating for penetrating finishes, because they do not form a coat.

- **Varnish.** See "Varnish Stain."
- **Polyurethane.** See "Polyurethane Stain."
- **Tung Oil.** No hardness rating after one coat. Soft to Medium hardness after two or more coats. A versatile finish which is both a penetrating and a coating finish. The first application soaks into the wood and successive applications coat it. Some craftsmen mix tung oil 1:1 with spar varnish to make a wipe-on 'penetrating varnish'. Tung oil can be used in all climates but it's too expensive for large projects.
- **Wood Sealer.** No hardness rating. A wood sealer penetrates deep into the wood, preserving it *and* sealing it against moisture. Contains a fungicide in a blend of vehicles. Preserves the look of raw wood and can be used in any climate.

Selecting the Appropriate Finish. As mentioned, you should select your exterior finish according to its hardness rating and the environment in which you live. Yet another consideration, just as important, is the *material* that the finish will be protecting. Use the following information as a guide in choosing the appropriate finish.

- Good choices for wood shingles and shakes include wood bleach, wood preservatives, and latex-base transparent and opaque stains. Paint is *not* usually recommended, unless it's a latex paint applied with an initial primer coat that includes a special 'stain block'. Often, shingles come factory-primed.
- Plywood may be stained with a light-bodied, semi-transparent oil base product if it is a top grade

PAINTING TIPS

- Use at least two brushes for your exterior painting job — a 4-inch one for flat areas and a 2-inch one for trim work and cutting in. A paint roller and tray will be useful for wide flat areas such as a soffit. Add an extension to the paint roller to reach high areas.
- Always use a quality ladder — one that is in good condition.
- Minimize your trips up and down the ladder — pour paint into a small bucket and hang it with a wire hanger from an extension ladder, or set the bucket on the fold-out support if you are using a stepladder.
- Set any exposed nails and use putty over the heads; this eliminates rust spotting.
- Protect your sidewalk, shrubbery, and flower beds by placing drop cloths over them.
- Stir the paint from the bottom up. Then, 'box' the paint. This technique involves pouring the paint from one container to another to achieve uniformity.
- In order to keep paint from dripping or running onto a freshly finished area, always work from the top down.
- Apply the paint in horizontal strips from your left or right, whichever is most convenient. After you've painted the peak of a wall, start at a corner and work across. Before moving, paint an area that measures about 4 or 5 feet square.

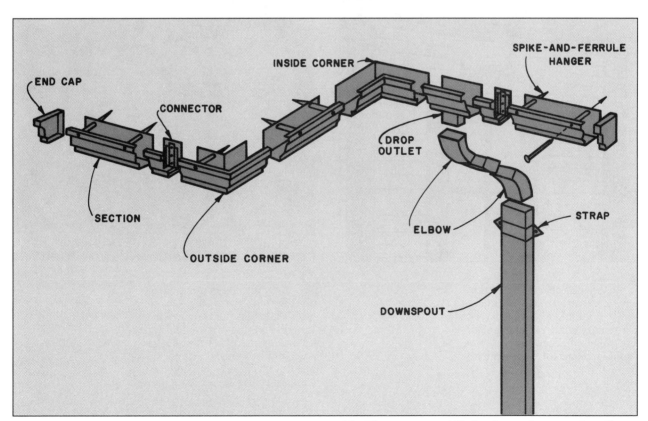

Components of a Gutter and Downspout System. A complete gutter and downspout system may be built from the many prefabricated components that are available. There are several kinds of gutter hangers; the spike-and-ferrule type is shown here.

ROOFING AND SIDING

119

plywood. Other grades are generally finished with a latex-base opaque stain, or with paint. (The *only* finish recommended for sanded plywood or plywood overlaid with resin is paint.) Some plywoods such as redwood, require a base coat of oil base primer or a specially formulated stain-blocking acrylic latex primer. These are finished with a top coat of a good quality acrylic latex paint.

■ Finish hardboard and hardboard panels with acrylic latex paints; be sure to apply an oil or acrylic latex primer coat first. Then, use *two* top coats. Be sure to check out the initial condition of the hardboard when you purchase it; it may be sold primed or completely finished.

Applying Exterior Finishes. Use the proper equipment and choose quality products to apply your exterior finishes. Paint may be applied by roller, brush, pad, or spray gun. A 4-inch brush is the standard size to use on wide surfaces. Choose a 2-inch brush for trim work and cutting in (where the painted surface meets a surface that is not painted or painted a different color). If you are painting large smooth surfaces, a roller with an 8-foot extension handle will be extremely helpful.

Use quality ladders or consider renting scaffolding equipment. Avoid painting during dusty, windy, or extremely hot and cold weather. The ideal temperatures for applying paint are between 50° and 90° F. If insects happen to land in your wet paint, brush them off after the paint dries. Refer to the 'Painting Tips' on page 119 for valuable information about preparation and application.

INSTALLING A GUTTER AND DOWNSPOUT SYSTEM

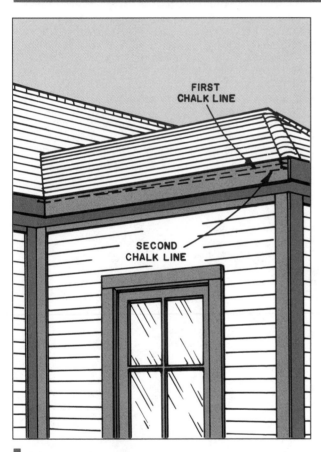

FIRST CHALK LINE

SECOND CHALK LINE

1 Insert nails at each end of the fascia and run a string from them. Check the string with a line level and then snap a chalk line. Using this line as a gauge, snap another chalk line running toward the downspout at a slope of 1 inch per 20 feet.

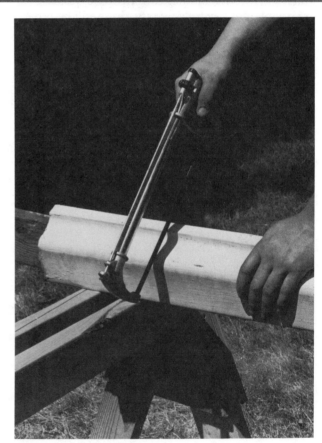

2 Assemble gutter components on the ground according to manufacturer's directions. If you need to cut a piece of gutter, place a 2 x 4 inside it for support and saw it with a hacksaw. Sand down rough edges with a file. If you're using spike-and-ferrule hangers, drill holes for the spikes.

The basic step-by-step process of applying paint with a brush is fairly simple: First, dip your brush about 2 inches into the paint and then tap off the excess against the inside rim of the paint can. Always coat the under edges of the siding. Be sure to 'feather' the ends of your brush strokes. This will join the painted areas with a minimum of brush marks and help eliminate runs.

Keeping the Structure Dry — Installing Gutters and Downspouts

Now that the roof and walls are completed, you need to protect your work with a well-planned gutter and downspout system (page 119).

Materials for gutters and downspouts run the gamut from galvanized steel and aluminum to copper or vinyl, but they can all be installed with just a few tools. When you're working on an add-on, try to match these 'accessories' with the rest of the house.

Gutters are designed to be attached to the fascia in a number of ways. Some have brackets or straps that are fastened to the fascia and then 'looped' around the gutter to support it. Other gutters are fastened by the spike-and-ferrule method which utilizes large spikes that are run through drilled holes at the top edges of the gutters and then driven into the fascia. Downspouts are simply attached to the siding.

The installation process is outlined below. Basically, assemble the gutters and then have a helper support the system as you fasten it.

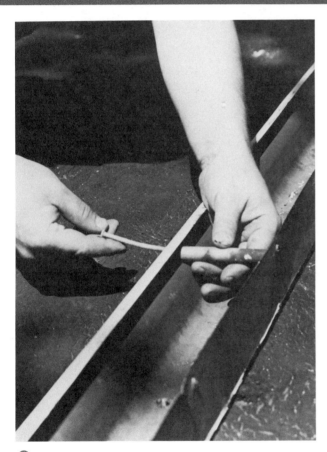

3 Prepare the sections of gutter for hanging. Here the spike-and-ferrule system is assembled.

4 Have a helper support one end of the gutter as you fasten the other end. Secure the gutter system to the fascia, working from the downspout inward. Next, assemble the downspout and attach it to the gutter; drill screwholes for connecting components to the drop outlet of metal gutters. Coat the back of straps with caulking compound. With straps and the appropriate fasteners, secure downspouts to the siding.

8

Finishing Touches

Finishing the interior of your structure is a very precise process because the work will be exposed — in clear view. An important step, however, is to locate and mark the stud and joist positions since practically everything is secured to these members of the framework.

No matter where you live, a northern or tropical climate, you'll need to add insulation. The types of insulation, the importance of vapor barriers, and how to install insulation are explained in this chapter.

Drywall is by far the most common wall covering. Cutting and fitting it is not too difficult, but it should be finished flawlessly. Sloppy seams will be visible through paint or wallpaper. Step-by-step instructions for installing drywall are given.

Some people prefer wall-to-wall carpeting throughout their homes. Others want vinyl tiles or the real thing — hardwood floors. If you're in the latter category, lay the floor or vinyl tile yourself by using the instructions in this chapter. You'll be amazed at how fast you can do it.

Finally, try your hand at installing trim work, molding, and door window casings. These accents, usually made of wood, are added where one surface meets another such as where the walls meet the floor (baseboards), the baseboard meets the floor (shoe molding), the ceiling meets the walls (crown molding), and the wall paneling meets at the corners. Also, learn how to install door and window casings.

Insulating the Shell

Now that you've moved indoors to continue your work, put your hammer aside and pick up a staple gun. The insulation must now be installed within the framework. This is critical for energy conservation, so don't skimp on materials. In fact, you should consult your local building department or other sources for information on just how much money you'll *save* by spending a little more on insulation.

The most popular types of insulation are fiberglass or mineral fiber which comes in rolls. This normally has a backing of foil or kraft paper which serves as a vapor barrier. Here are some guidelines to follow when installing insulation. As a preliminary step, you can use a special 'filler foam', as shown below, that fills in holes in the sheathing created by the electrician or plumber. When you install insulation between the studs, place the vapor barrier facing inside the structure (page 125). If you won't be using your attic as a living space, lay insulation between ceiling joists with the vapor barrier facing downward. When you're insulating between floor joists over an unheated area, lay it with the vapor barrier facing upward. If the type of insulation you're using doesn't come with a vapor barrier, use a separate vapor barrier in the form of plastic sheeting.

Typically, insulation is installed by stapling the paper flanges on the insulation to studs and joists. Ends should be left a little long and then tucked in for a tight fit. For special situations, other methods of attachment are used — such as wire braces or baling wire.

Although energy consciousness has its rewards, do not put insulation everywhere. Leave attic vents and chimney flues as they are, and also do not cover *anything* that produces heat, such as fans and other electrical equipment.

Doing Your Own Drywalling

Although there are several choices in wall coverings, such as boards and wood paneling, the most accepted one for practicality and versatility is *gypsum wallboard*. Widely referred to as *drywall*, this sheeting material is composed of chalk-like gypsum and softening agents — primarily shredded paper. The substances are pressed between two backing sheets of heavy paper.

Although 1/2-inch drywall is most commonly used, it is also made in 3/8- and 5/8-inch thicknesses. The standard panel width is 4 feet, but 8-, 10-, and 12-foot panels are also available. Choose the longest panels that you can handle comfortably.

Screws, nails, or a combination of nails and adhesive may be used for the fastening process. Nailing is the most popular method. Nail size must match panel thickness. For 3/8- and 1/2-inch panels, buy 1 1/4-inch nails. For 5/8-inch panels, use 1 5/8-inch nails. Specify ring-shank nails with 1/4-inch-diameter heads.

Basic Installation Methods. Drywall nails and screws are designed specifically for insertion into this material. A special nailing technique must be used. Nailheads should be driven below the surface of the drywall to slightly 'dimple' it — but they should not be struck so hard that the paper is dented. The use

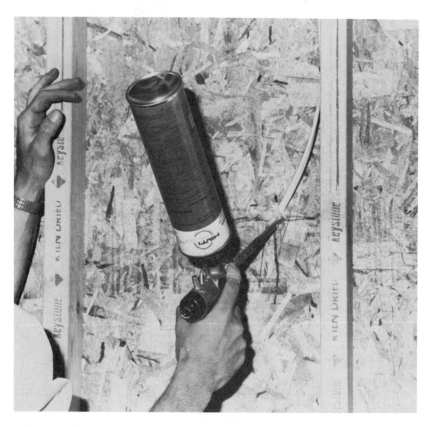

Filler foam is used to plug up holes in the sheathing that were made for the plumbing and wiring systems.

of a bell-faced hammer will make this job easier. Screws likewise should be installed just beneath the wall surface. In general, first install a few fasteners in strategic places to hold a panel in place. Then 'fill in' with the rest of the fasteners.

Spacing is determined by local building codes; typically 1/8 inch is sufficient, both at seams and at intermediate studs or joists (the area referred to as 'in the field' by builders). Position fasteners a minimum of 3/8 inch from panel edges.

For extra holding power, 'double-nail' by driving extra nails placed 2 inches from the originals. If this method is used, the pairs may be spaced only 12 inches apart in the field. The edges of drywall panels are slightly tapered on their fronts so that, when joined, they form a subtle recess for building up the seam with joint compound.

Drywall has a certain lightness and flexibility that makes it fairly easy to work with. On the other hand, these qualities make it a fragile material. You must handle it gently to avoid damage at corners and ends. Use a metal T-square and a utility knife to mark and cut it. Mark and cut it on one side and then snap it along the scored line. Turn the sheet over and cut the paper on the other side. Smooth rough edges with a perforated rasp.

INSTALLING INSULATION

1 There are several types of insulation and methods by which they may be installed. Pressing fiberglass insulation (with a vapor barrier) into place between the studs is shown here. Note that the vapor barrier faces toward the inside of the structure. The insulation is stapled to the studs.

2 If you are using insulation without a vapor barrier attached, staple plastic sheeting to the studs to act as the vapor barrier and to hold the insulation in place.

CAUTION Because all insulation materials shed and may be irritating to your skin, eyes, and breathing passages, you should protect yourself when applying them. Always wear a face mask or respirator, a hat, a long-sleeved shirt, goggles, and gloves.

Panels may be installed either parallel or perpendicular to framing members. The ceiling panels will be the most difficult to install. You'll need several helpers to lift the panels into place. At studs, install panels vertically unless your walls measure higher than 8 feet. Horizontal application in this case would create too many seams.

Use the step-by-step instructions on this page and the next for basic drywall installation. Then for taping joints, see pages 128 and 129.

After the joint compound dries, you may want to paint or wallpaper before the moldings and floor are installed.

Repair Methods. While you've got the materials, the joint compound and the tools all handy, you might want to inspect your home for any damaged drywall. Making and installing a patch is quite easy (page 130).

If you have an older home which has walls composed of lath and plaster, the same materials may be

INSTALLING DRYWALL

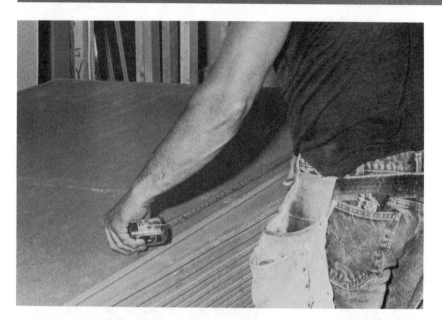

1 To measure for drywall, first take measurements from the wall you intend to cover. Transfer these dimensions to the drywall. Hold your thumb on the measurement of the tape and hold your pencil in the other hand; then, simply pull downward to make the long pencil mark. Use the same basic method when measuring for any openings in the panel, such as those needed for electrical outlet boxes.

3 You'll need one or more helpers and several ladders when you install the ceiling panels. Raise each panel and support it with your hands. Install the first few nails near the center of the panel. Then finish the nailing according to your local building code.

FINISHING TOUCHES

used, but the process is a bit trickier. In order to ensure that the new patch will be level to the old wall, you'll have to shim furring strips to framing members and attach the drywall to them. This process is shown on page 131.

Finish Floors

In home building terminology, the 'finish floor' refers to the final covering placed on the flooring.

This surface should be chosen according to the use of the room. For basements, garages, and workshops, the final flooring can be simple concrete, or, if desired, concrete with a heavy-duty paint for sealing out stains and grease.

Entryways can be surfaced with ceramic tile or similar materials such as flagstones or marble. These give your framing the undisputed best protection from moisture. For other 'wet' areas, such as kitchens and play areas, synthetic flooring is often chosen.

2 Cut the panels with a utility knife. Cut small openings with a keyhole saw.

4 Install the wall panels either vertically or horizontally. Vertically is recommended because the panels will be easier to handle. Shown is a vertical installation. If your walls are higher than 8 feet, install the panels horizontally. However, be sure to stagger the end joints so vertical joints in the top and bottom rows don't match. Center all edges over wall studs and nail according to your local building code.

5 Be sure to hammer the nails with just the right amount of force to 'dimple' the drywall material. A special bell-faced hammer is used.

FINISHING TOUCHES

1 Using a joint knife, spread a smooth layer of premixed joint compound over all nail holes.

2 Smooth the joint compound and make the surface even.

CAUTION Whatever type of joint compound you use, be sure to follow the manufacturer's directions and cautions.

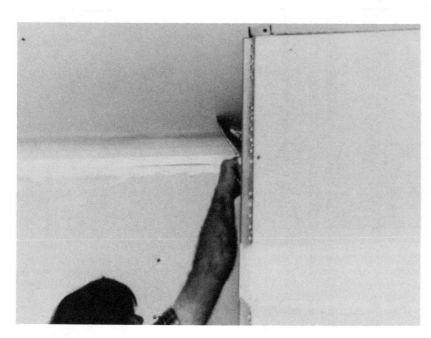

5 For outside corners, nail on a metal *cornerbead* and spread joint compound over it.

3 At drywall joints, spread on a layer of compound. Before the compound dries, press a strip of pre-creased, perforated drywall tape into it.

4 Use the joint knife to smooth out and remove excess compound. For inside corners, use a special corner tool.

6 Allow the joint compound to dry completely — approximately 24 hours. Wear protective clothing — preferably a mask, goggles, and a hat — for the sanding. Sand joints until smooth. Use a 10-inch knife to apply another coat of compound; feather the edges beyond the previous coat. Allow it to dry, then smooth it with fine sandpaper.

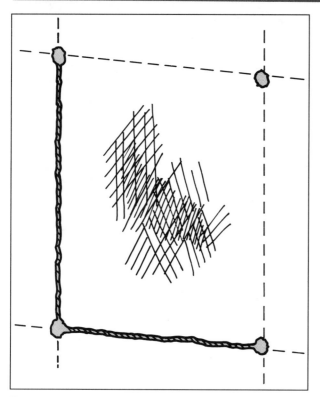

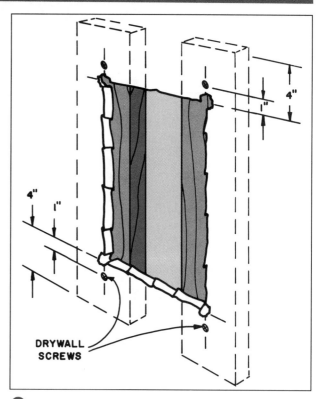

1 With a square, draw a rectangle around the damaged section as shown. Drill holes at each corner and cut the section out with a keyhole or drywall saw. Carefully pull the section forward as you finish the cut; do not let it drop behind the wall. Use this as a pattern for cutting a new piece of drywall of the same thickness.

2 Cut two 1-inch thick boards to fit at both sides of the opening; they should extend approximately 4 inches above and below it. Install the boards, centered at the sides, as shown. Use drywall screws for fastening them inside the wall; place the screws 1 inch above and 1 inch below the opening, and, if the hole is large, at 8-inch intervals on the sides.

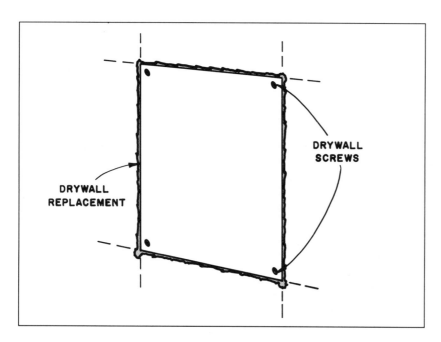

3 Insert the new piece of drywall into the opening and anchor it to the boards with drywall screws at each corner, and if needed, at 8-inch intervals. Cover the seams and screw holes with drywall tape and joint compound. Feather the joint compound to make the pieces of drywall even. Sand the joint compound smooth.

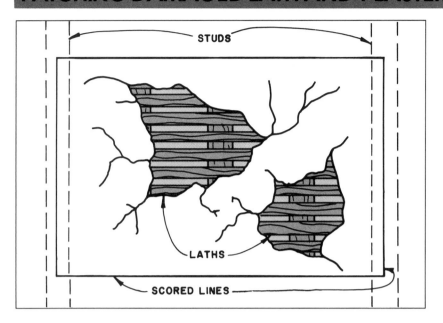

1 Wear protective eyewear and clothing. Use a square to draw rectangular lines that encompass the damaged plaster. Center two of the lines on ceiling joists, or, in the case of walls, on studs, and draw the other two lines at right angles. With a utility knife, score the lines. Using a cold chisel and hammer, gradually chip away the plaster, moving from the center to the perimeter. Work carefully, chipping small pieces. If you make small cracks outside the rectangular area, you can patch them. If any plaster outside the rectangular area becomes loosened from the lath, extend the rectangular area.

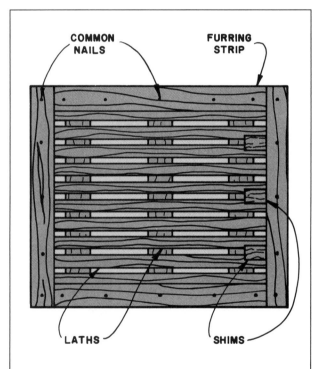

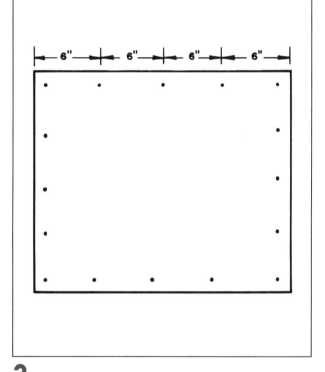

2 Cut furring strips to fit the perimeter of the rectangular area. Using small common nails, nail the furring strips partway into the ceiling joists (if a ceiling patch), the studs (if a wall patch) and the lath. With a small piece of drywall, check each side to be sure that the plaster surface will be raised slightly higher than the new piece of drywall. If necessary, insert shims under the furring strips. With the shims in place, finish driving the nails.

3 Cut a piece of drywall to just fit the rectangular area. Drive a nail at each corner through the drywall, furring strips, lath, joists (if a ceiling patch), and studs (if a wall patch). Also drive nails at 6-inch intervals around the perimeter as shown. Cover the nails and seams with drywall tape and joint compound. Feather the joint compound and sand smooth to make the drywall even with the plaster.

Vinyl and a host of similar types of flooring, in both sheets and individual tiles, are available under many brand names.

Some vinyls are also referred to as 'resilient' flooring because they provide a slight cushion for your feet. Because they are relatively thin and soft, when you use resilient flooring, you should first install an *underlayment*. This layer of flooring, in the form of hardboard, plywood, or particleboard, will cover the irregularities common in the subflooring. Install resilient flooring according to manufacturers' specifications.

For living areas, a modern trend is to use wall-to-wall carpeting wherever possible. When carpeting and the pad are laid directly over subflooring, the cost of putting in a finish floor is eliminated entirely.

Strip Flooring — Long on Wear and Beauty

Nothing can beat the attractiveness and warmth of wood flooring. Moreover, wood floors are relatively hard, and yet they have a fair degree of resiliency.

Wood floors can be installed in several forms — strip, plank, parquet, and block — but the most popular kind, and the easiest to work with, is strip flooring.

Floor strips are manufactured with tongue and groove joints. Each strip is shaped to be joined on both its sides and its ends. Although oak is the most preferred, other hardwoods are available; birch, maple, hickory, beech, and ash all can be used. Softwoods frequently chosen are fir and yellow pine.

Strip flooring is rated by quality according to grain, color, and number of imperfections. The grades are Clear, Select, No. 1 Common, and No. 2 Common — in order of decreasing quality. Boards are sold in random lengths from 9 to 102 inches. Although strips come in various dimensions, the most commonly used are $^{25}/_{32}$ inch thick by $2^{1}/_{4}$ inches wide.

Purchasing and Preparing. Strip flooring is sold by a 'board foot' formula which is based on pre-milled sizes. When you order the material from your lumberyard or flooring distributor, the information that they'll need is the area that you're going to cover — expressed in square feet.

LAYING WOOD FLOORS

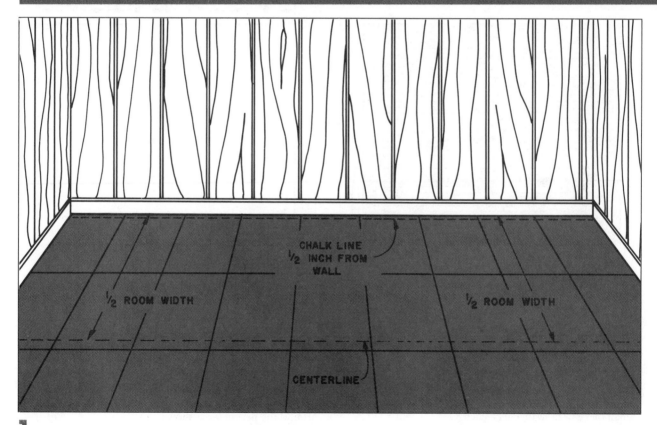

1 Mark the starter course with a chalk line.

When your flooring is delivered, stack it in the room where it will be installed. Adjust the room temperature to normal living conditions; that is, heat it if it's cold, and run an air conditioner if it's too hot. Remove flooring from its packaging and let it adjust to the room's humidity and temperature for at least three days.

Practically everything else should be done before the finish floor is installed; this includes plumbing, wiring and all wall and ceiling surfacing such as painting and papering. The only thing that *shouldn't* be done yet is the baseboard molding.

One step you'll need to take before you do the actual nailing is to lay down 15-pound building paper. If your subflooring is made of boards, this will give protection at the board seams. If you have plywood subflooring, the building paper is not truly needed for protection against moisture, but it is highly recommended as a good sound barrier.

As you lay down the building paper, be sure to mark it with chalk lines for the floor joists so you'll know where to fasten the flooring. This pattern should be visible on the subflooring. If not, you'll need to drill pilot holes to locate the joists.

The nailing will be done primarily with a special rented tool, an electric *flooring nailer,* which machine-drives cleats into the groove side of the floor strips. With it comes a hefty 5-pound rubber-headed mallet for striking the machine's head. This machine will be used on the expanse of the floor except at the ends where it will not fit.

Measuring and Marking.   In order to determine the starter course, draw a centerline across the room in the same direction that you plan to run the flooring. Preferably, floor boards should be run perpendicular to floor joists. Measure exact distances from both ends of this mark and snap a chalk line at the end of the room where the first course of flooring will be laid — about ½ inch from the wall. Doing this ensures that the middle of the floor will have square, true strips. The edge strips may not be equal distances from the walls, but any irregularities can be disguised with molding.

Read the step-by-step instructions for laying a wood floor, starting on page 132. Then practice using the flooring nailer a few times. Begin by laying out six to eight courses of flooring at a time, fitting them like puzzle pieces.

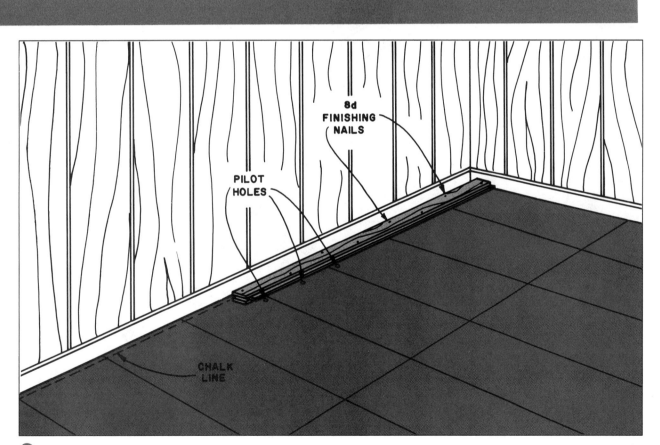

2 Apply the first course of flooring by face-nailing it, tongue out, along the chalk line. Do this by drilling pilot holes for 8d finishing nails at each end of the board as near as possible to the grooved edge. Drive nails here and also at joists and between joists. Shoe molding will cover the nail holes.

3 Lay out about eight rows of boards so they fit within end walls with less than ½ inch slack. Stagger the end joints in adjacent rows so they are at least 6 inches apart. Push the boards snug.

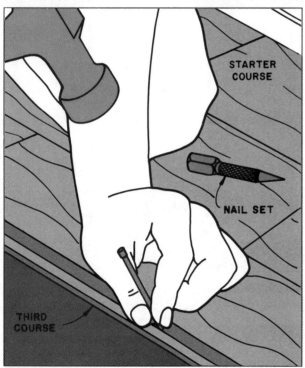

STARTER COURSE

NAIL SET

THIRD COURSE

4 Drive and set nails in the second and third rows — an area too tight for the flooring nailer. Blind-nail the boards by driving nails at a 45° angle through the corner of the tongue, as shown.

FLOORING NAILER

5 Fit the flooring nailer over the tongue of the board and strike its plunger with the rubber mallet. Hold each board in place with the heel of your foot. Drive in cleats about two inches from each wall, at every joist, and halfway between joists. If a cleat fails to penetrate, pry it out with a claw hammer.

Easy-to-Lay Vinyl Floor Tiles

Floor tiles offer homeowners an easy and fast way to install flooring. Whether the floor tiles are vinyl, cork, rubber, or wood parquet, the process is much the same. Mark up a plan on graph paper, create guidelines on the floor, run a 'dry' or preliminary course, and then begin installation.

The commonly available sizes of floor tiles are 9 and 12 inches square. Wood parquet flooring, on the other hand, comes in many different widths and lengths. For example, one manufacturer offers 6-inch squares while another sells 'slats' measuring 4¾ inches wide by 19 inches long.

If you're installing tiles on a slab, be sure the concrete is flat and smooth. Fill or flatten any spots that need it. An epoxy filler sometimes called 'flash patch' is the ideal material for this step. If you need to flatten bumps, use a rented electric concrete grinder. Coat the surface of the concrete with a sealer.

Your supplier should help you determine how many tiles you need if you give him the dimensions of the room. If you're tiling an irregularly shaped room, take a drawing of it with you when you go to order the flooring. Most dealers will gladly accept returns if they've overestimated.

A newer, and very popular flooring product is 'self-adhesive' vinyl tiles. These are very easy to use since the step of spreading out the adhesive is eliminated. In fact, the most difficult thing about laying these tiles is removing the backing on the tiles. Follow the instructions on this page to install self-adhesive vinyl floor tiles.

LAYING SELF-ADHESIVE VINYL TILES

1 After the floor has been checked for level and the sealer applied and allowed to dry, execute a dry run of one row of tiles to see how they will fit in the floor space. It's best to 'center' the tiles within the space and cut the end tiles to fit. Start to install the tiles. Remove the paper backing and carefully lay each tile into place, butting it against the previously installed tile.

2 When you're satisfied with the position of the tile, apply pressure at each seam — using the heel of your hand or a roller. Then smooth your hands across the middle of the tile. Whenever you need to cut tiles, use a utility knife. You may want to wear knee pads if you're installing tile in a large room.

CAUTION If your tile is the type that requires an adhesive, follow the manufacturer's instructions. Most adhesives are toxic and flammable, so keep them away from open flame and keep the room well ventilated.

1 You can apply stain or paint to the molding before installation.

2 For an *outside corner,* cut both pieces at a 45° angle. If they don't match perfectly, trim the pieces at the back with a utility knife and at the front with a small block plane.

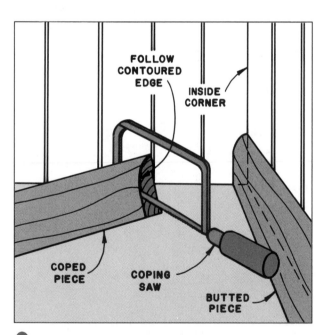

FOLLOW CONTOURED EDGE

INSIDE CORNER

COPED PIECE

COPING SAW

BUTTED PIECE

3 Coping is the method used to fit shaped moldings that meet at an *inside corner.* First measure and cut both pieces. Leave the 90° cut on one of the pieces and cut the other at 45°. On the piece that was cut at 45°, place the coping saw at 90° to the back edge. Make the cut following the contour. This coped joint should fit perfectly against the first piece.

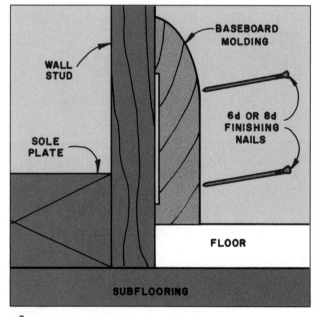

BASEBOARD MOLDING

WALL STUD

6d OR 8d FINISHING NAILS

SOLE PLATE

FLOOR

SUBFLOORING

4 Nail molding with two 6d or 8d finishing nails or color-matched nails positioned at each stud location. Fasten the top nail to the wall stud, the lower nail to the sole plate. Set the heads of the nails. If you're installing crown molding, nail it to the wall studs and top plates.

Final Touches — Trim and Molding

The finale of your interior work, after everything else is installed and finished, is the process of adding trim and molding. These are used to hide the framework at unsightly seams, to protect surfaces, to complete the look of the room, and to add a decorative touch.

The term *molding* refers to strips of wood or other nonwood products with shaped surfaces and edges that are applied at the base of the walls *(baseboard molding)* or the place where walls meet ceilings *(crown molding)*. *Shoe molding,* in the form of quarter-round, further connects floors to baseboards. Other types of trim and molding are used for chair rails, coves, cornices, and at the corners of wood-paneled walls.

In these exposed spots your woodworking will be on display so you'll want to take extra care with the fine details. Basically, what's required is very studied measuring and cutting plus careful nailing. It's usually best to do any staining or finishing of the trim *before* it's applied but there may be circumstances when it's necessary to do the staining after installation.

For molding and trim, you'll be cutting mostly 45° angles with a quality miter box and backsaw, or a precision power saw. If you need to make unusual angles, you might consider renting a power miter saw. A sharp utility knife or a block plane will be useful for trimming away excess material. In order to install molding at inside corners, you'll need a coping saw to make allowances for the curvature of the adjoining piece. This special process is called *coping a joint*.

Working with Moldings. Before you start, make sure that you have transferred markings for stud positions — so you'll know where to do the nailing. To install the base molding, start at the longest wall. Measure and cut for a snug fit and then tap the molding into place with a block of wood. If you can't find a length of molding that is long enough, you can simply miter two pieces together.

There are two choices for nailing the molding into place. Use either color-matched nails or regular finishing nails which will need to be recessed with a nail set. At outside corners, use miter joints to connect moldings; at inside corners, make coped joints. Molding at the crown is cut and installed in the same manner. Drive nails into the top plate, or, if the molding is heavy, into the ceiling joists as well.

Shoe molding is the last thing to be installed. Wait until windows and doors have been trimmed and also after the floor has received its final covering. Cut and fit shoe molding in the same manner as above. Nail it to the baseboard, not the floor. Where the shoe molding meets a door casing either round it or miter it for a clean look. Instructions for installing molding are found on page 136.

Trimming Windows and Doors. In Chapter 7, you were given the instructions for installing windows and doors. Now that you're doing the interior finish work, you may need to add door and window casings. There are two methods. Either use solid board lumber with butt joints, or, for a more standard and professional look, use miter joints.

The process for installing door and window casings is shown on pages 138 and 139.

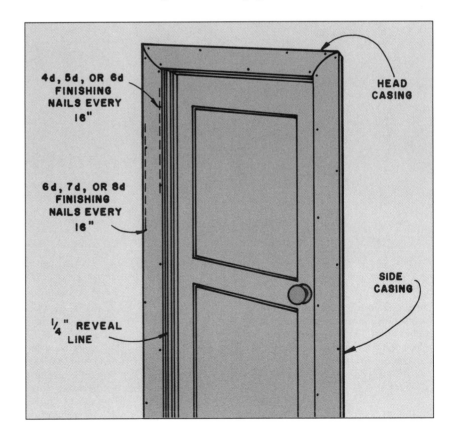

Interior Door Casings. This interior door unit includes head and side casings. After marking the reveal lines you'll measure, cut, and install the head and side casings.

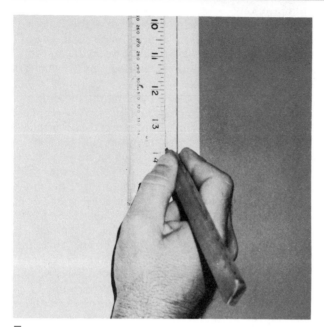

1 Mark a setback line or *reveal line* ¼ inch inside the edge of each door jamb. Align the head casing with this line and mark end points. From these points, cut 45° angles.

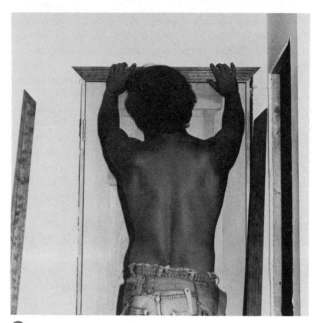

2 Check the head casing to make sure that you have cut it correctly.

3 Attach the head casing to the jamb at its inside edge with 4d, 5d, or 6d finishing nails; space nails 16 inches apart. Also fasten the casing to the rough framing with 6d, 7d or 8d nails spaced 16 inches apart.

4 Measure for the side casings and cut the 45° angles at the tops. If the finish floor is not in place, allow for it at the bottom. Install the side casings using the same nailing method as for the head casing. Also toenail through the head casing and side casing into the door jamb. Then set all the nail heads in the side and head casings.

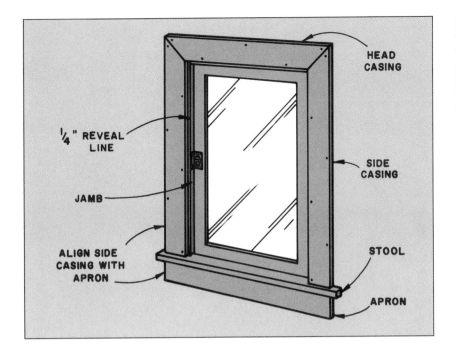

Interior Window Casings. This window unit includes *side casings,* a top or *head casing,* a bottom casing or *apron* and a *stool.* Install the components in this order: stool, side casings, head casing, and apron. Use finishing nails for all the fastening.

INSTALLING WINDOW CASING

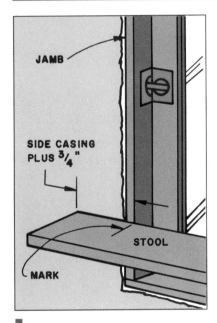

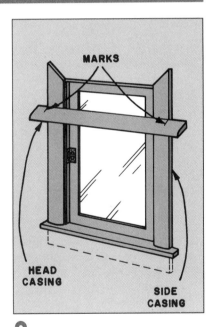

1 Begin by marking the ¼ inch reveal lines (see "Installing Door Casing"). Measure and mark the width of the casing plus ¾ inch out from the reveal lines on each side of the window. The distance between the marks is the length of the stool. Cut the stool, then mark the inside edge of both side jambs on the stool's back side.

2 Place the stool as shown. Make marks where the stool meets the outside edges of the jamb and the sash. Repeat with the other end of the stool. Use a square to intersect the lines for the end notches. Cut the notches. Then fasten the stool to the sill with 6d nails. Measure for each side casing by setting a length of molding on the stool and marking where it meets the head casing. Cut 45° angles. Use 4d or 6d nails to attach casings to the jamb and 8d nails to attach them to the rough framing.

3 Measure for the head casing by holding a length of molding to the outside edges of the side casings. Cut one end at 45°. Test it for a snug fit; then re-measure and cut the other end at 45°. Make adjustments as necessary. Use the same nailing method as for the side casings. Measure for the apron; the ends

Home Cabinetry

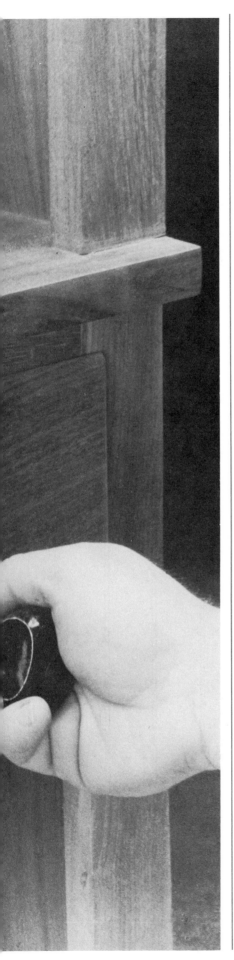

9

Basic Cabinet Construction

You've just completed an add-on to your home. Maybe it's a family room and you're now having visions of it with built-in cabinets and bookcases. Or it's a kitchen that needs cabinets. In both of these situations, custom built-ins will give an uncommon look to the room. Why? Because they're precisely tailored to fit and they're constructed of materials that will suit your taste and match your decor.

The selection of materials for built-ins is broad. If possible, look for inspiration at finished cabinets and bookcases at a home improvement center, builder's supply store or cabinetmaker's shop. Try to find out as much about them as possible...how they were made, and what woods and finishes were used.

This chapter will give you the basic knowledge you'll need to design and build your own cabinets and bookcases. Here you'll see the logical progression of cabinet construction. Every component is covered including carcases, drawers, doors, and shelves. Most importantly, you'll inherit a feel for wood and learn to join it according to grain pattern. The topic of finishes is also presented; most of the basic types of finishes are discussed so that you'll have a good idea of which will be the most appropriate for your built-ins.

Cabinet hardware, such as drawer pulls, door handles and door hinges is discussed in this chapter. Shelves can be mounted with the use of metal standards and another form of hardware, drawer glides, are sometimes chosen over wooden guides for supporting the drawers. Installation tips are presented for all these.

As a beginner, you're bound to make a few errors. If you damage your attractive work surface, don't despair. Use the repair instructions toward the end of the chapter to make the wood right again. For now though, concentrate on the positives. Sit back, read and absorb these cabinetmaking principles. Study the project plans thoroughly. Then get out your measuring tools and drawing pencil. It's time to create your own built-ins.

Cabinetmaking — Step-By-Step

There are many different methods to use to build cabinets and obviously thousands of plans to choose from. As a beginner, you should stick with very simple plans, so your cutting and measuring will be straightforward and uncomplicated.

You might want to design your own cabinets. When drawing your plans, it's best to use the industry standards that are set for cabinet dimensions (shown below). Designed in this way, your cabinets will be sized to fit with appliances such as a range or dishwasher, or match any cabinets that you might already have in your home.

A word of advice: Before you begin, have a clear plan for every part of the cabinet — down to the last screw. Figure out the type of carcase, what joints to use, the style of door, how the drawer will glide in and out of the framework, how shelves will be hung and what type of hardware will be used.

Another consideration in building cabinets is the weight of the finished structure. This factor determines your use of materials, joints and fasteners. Determine what finish you'll be applying. The type of finish will determine the materials you'll use. If you plan to stain the cabinet, use wood or hardwood veneer plywood on the surfaces that will be visible. If the side(s) will be invisible or if you plan to paint the

STANDARD DIMENSIONS FOR CABINETS

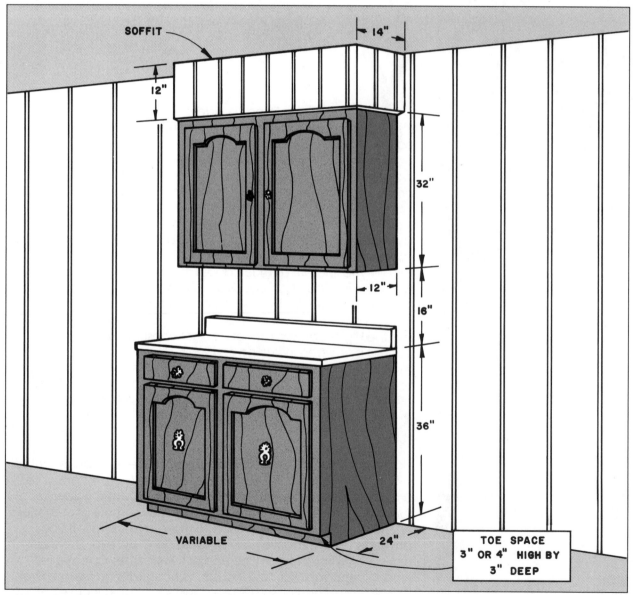

These are industry-wide standards for kitchen and bath cabinets. The dimensions should be adhered to although you may vary heights slightly to suit your needs.

cabinet, you can use less expensive grades of wood or plywood.

You can make allowances for all of these factors early on in your measuring, cutting and joining, but you must have a plan before you start.

There are many different approaches to building cabinets, but there is one thing that you must pay close attention to — the direction of the wood grain. This will affect the overall look of the finished cabinet. Cabinet parts are joined with grain direction in mind and appropriate joints are used to allow for wood movement. For a refresher to this procedure, review page 41.

After you've drawn your plans and selected your materials, it's time to get started. Here is the most logical progression for building one cabinet or a whole roomful of them:

1. Build the carcase.
2. Add facings and shelves.
3. Build and install drawers.
4. Build and install doors.
5. Apply the finish.
6. Add pulls and handles.
7. Mount the cabinet to the wall.
8. Add the cabinet top.

☐ **WOODWORKER'S TIP** ☐

If you are building several cabinets with the same or similar components, cut all the primary parts — sides, bottoms, drawer components, doors, face frames, toe kicks and shelves — for all the cabinets first. Remember the rule: 'Measure twice, cut once'. Next cut the joinery and add the decorative touches. Then assemble the components. By using this method, you'll save time setting up your power tools, especially if some of the cabinets are identical, plus you'll generally have a more orderly operation.

Building the Carcase

The main component of any cabinet is called the carcase. Other components such as drawers, shelves, and doors attach to it. The three types of carcase construction — case, frame, and panel — are shown on pages 146 and 147. *Case construction,* considered the simplest type, is merely panels joined with simple joinery. *Frame construction,* similar to the framing of a house, consists of a framework that is covered with wood, plywood, particleboard or hardboard. *Panel construction* is an elaborate variation of frame construction. Instead of attaching the panels to the outside of the framework, they are held in grooves cut in the framework. This type of construction allows for greater expansion and contraction of the wood.

Adding Shelves

Shelves, important components of most cabinets, can be fixed or adjustable. One type of *fixed shelves* has simple wooden cleats that are fastened to the carcase sides. The shelves are laid on or fastened to the cleats. Another type of *fixed shelves* can be used if you are working with case type construction. These shelves slide into dadoes that are cut in the carcase sides. Fixed shelves will add additional strength to the cabinet.

Adjustable shelves, on the other hand, add little strength to the cabinet, however they can be moved to accommodate various-sized objects. One method of installing *adjustable shelves* is to drill a series of holes in the sides of the carcase and add dowels or special shelf holders to support the shelves.

The most popular method for installing *adjustable shelves is accomplished with the use of metal or plastic* shelf standards. These standards allow the shelves to be moved quickly, usually with just the touch of a finger. Please note, it's best to install standards before the carcase is assembled; they may, however, be put in later. Follow the instructions on page 148 to install fixed and adjustable shelves.

Making the Drawers

Many beginners are intimidated by the complexity of drawers and are therefore reluctant to tackle them. However, as shown on page 149, the most basic drawer is an 'easy-to-make box' that can be built quite easily in a few simple steps. This 'box' is not too attractive, but it helps to illustrate just how simple drawer construction can be.

There are many different joints and combinations of joints that can be used for drawer construction. The components and joinery can be as simple or elaborate as you want. Basically drawer-making involves assembling the 'box' with four framing members and then adding a bottom to it in some fashion. The illustration on page 149 shows typical drawer construction. The drawer may be flush with the carcase, lipped, or overlapped as shown on pages 150 and 151.

A drawer is subjected to a great amount of stress, primarily on its front, from the pulling it receives by users. Therefore, the most important joints are those that connect the sides to the front. An easy-to-make, strong joint to use when attaching the front to the sides is the *lock joint.* Use the instructions on pages 150 and 151 to make a lock joint.

The joints used at the back of the drawer are not as important — first, because they don't receive as much stress, and second, because they are not visible. An easy design choice here is simple dadoes.

The final step in constructing the drawer is to add the bottom. Since this is the largest part of the drawer, it is subject to the most distortion when the temperature or humidity changes. For this reason, hardboard or thin plywood is often chosen as the appropriate material. If solid wood is used, it should be a close-grained variety such as poplar or pine.

The three basic types of carcase construction — case, frame, and panel — are shown here.

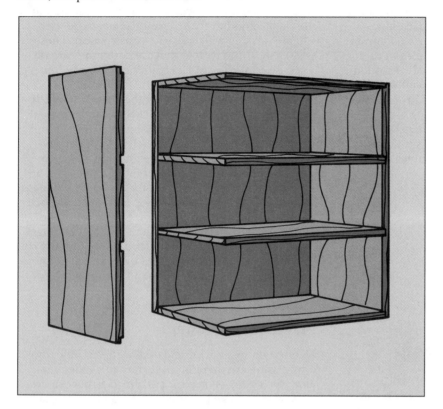

Case construction, the type with simple joinery and the least number of components, is usually the easiest and fastest to build. Plywood and particleboard are the best materials to use. The shelves slide into dadoes that are cut in the sides, and the back fits flush against the shelves and sides. The top and bottom fit into rabbets cut into the top and bottom of the sides.

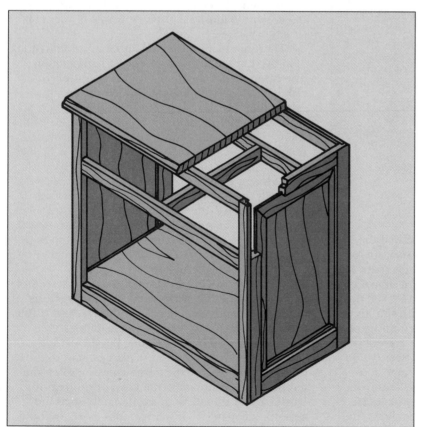

Panel construction is an elaborate and more decorative type of carcase. The side panels are held in place by grooves cut into the stiles and rails. The panels are unglued and floating to allow for expansion. Dadoes, rabbets and mortise and tenons are used to join the components.

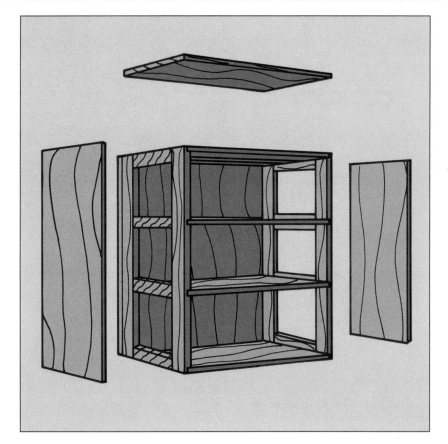

Frame construction is similar to the framing for a house. It's a framework of inexpensive lumber that is covered with wood, plywood, or particleboard. Therefore it is quite sturdy. The side frames are built first, then the dadoes and rabbets are cut to accommodate the shelves and the back. The framework is assembled with dowels, lap joints, or mortise and tenon joints.

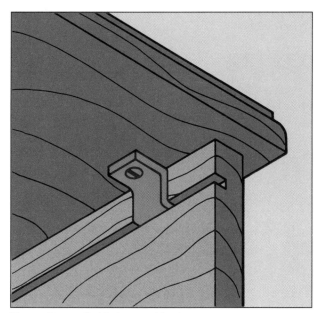

Top. Generally kitchen and bathroom cabinets have plastic laminate countertops. Hardwood veneer plywood or glued-up wood are used to top other types of cabinets. Tops usually attach to the frame with metal brackets as shown here or fit into rabbets cut in the tops of the carcase.

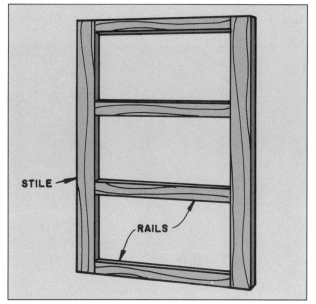

Face Frame. Add a face frame to the front of the carcase to cover the exposed edges, to hide framing imperfections and to provide a solid surface for mounting the door. Cut butt or miter joints, and join them with dowels or mortise and tenons. Attach the face frame to the carcase with glue or finishing nails.

BASIC CABINET CONSTRUCTION

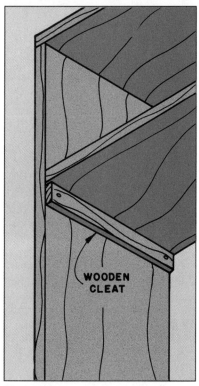

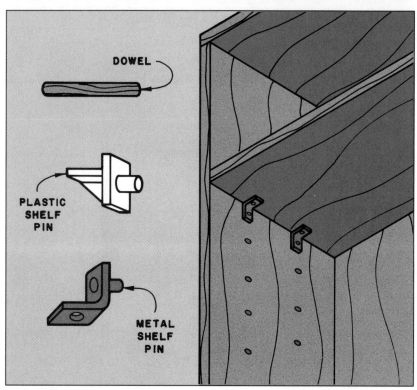

Fixed shelves can be held in place with *wooden cleats* that are mounted to the carcase sides. To determine the length of the shelves, measure between the sides of the carcase and subtract ⅛ inch to allow for installation.

One method of installing **adjustable shelves** is to drill vertical rows of holes in the carcase sides. *Dowels* or special *shelf pins* are inserted in the holes to support the shelves. In the case of dowels, the underside of the

shelf where it sits on the dowel should be grooved slightly.

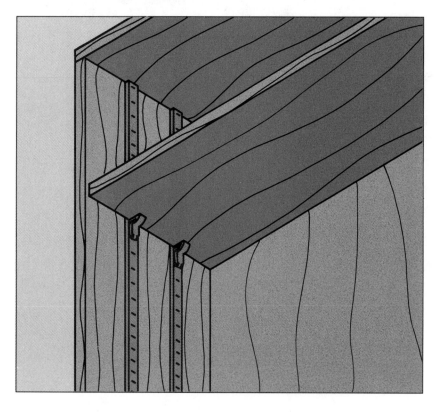

The most popular type of **adjustable shelf** is accomplished by the use of metal or plastic *shelf standards*. First read the hardware manufacturer's directions to determine if the standards have 'up' and 'down' ends. Note: It's best to mount the standards to the sides before assembling the carcase. Mark the carcase sides, 1 inch from the back and 1 inch from the front. Hold a standard along the mark. Drill pilot holes and attach the standard with screws. Repeat the process for the remaining standards. Be sure the tops of all the standards are aligned so the shelves will be level. Measure the distance between the installed standards and cut the shelves ⅛ inch shorter than this measurement.

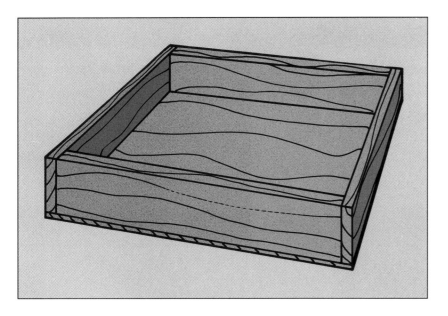

A Butt Joint Drawer. You can make a *very* simple drawer by using all butt joints. Reinforce the joints with nails or screws and use a piece of ¼-inch plywood or ⅛-inch hardboard for the bottom. Although this drawer is strong enough for light use, it is limited in strength and it's not too attractive.

TYPICAL DRAWER CONSTRUCTION

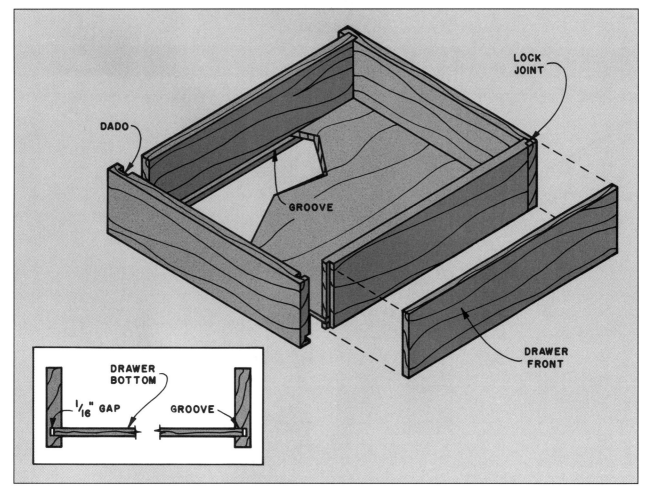

This drawer is well constructed, and is relatively easy to make. The sides are connected to the back with simple dadoes and a lock joint is used to hold the drawer front to the sides.

The drawer bottom is fitted loosely into grooves cut in the drawer front, back, and sides. The bottom should be cut ⅛ inch 'short' on each side to allow for a ¹⁄₁₆-inch gap in each

groove. The gap should be slightly larger if the drawer bottom is made of solid wood.

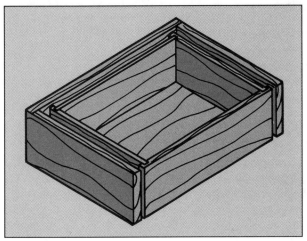

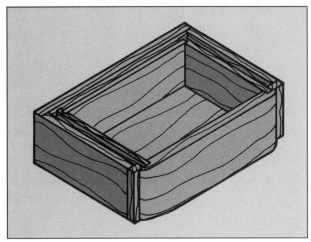

A flush drawer is precisely even with or slightly recessed into the carcase. The carcase contains an interior 'stop' that prohibits the door from sliding further inside when it is closed.

A lipped drawer has a front that is rabbeted so there is a ¼-inch to ½-inch lip all around the outside edge of the drawer. This part of the drawer front functions as a stop when the drawer is closed.

MAKING A LOCK JOINT

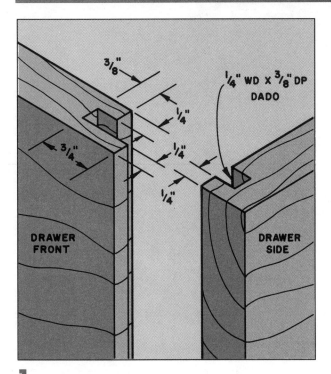

1 The lock joint is very strong and easy to make. Use ¾" stock for the drawer front and sides.

2 Cut a ¼" wide x ⅜" deep dado on the inside surface of the sides.

CAUTION Always wear proper eye protection when using power tools (pages 3 and 4). Also, be sure to keep your hands out of the danger zone (page 5).

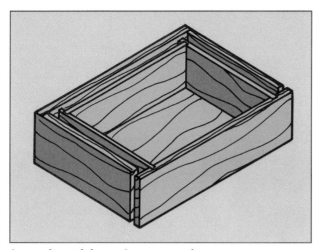

An overlapped drawer is constructed with the entire face wider and/or taller than the drawer opening. Thus a stop is provided.

The drawer bottom is joined to the sides with classic frame-and-panel joinery. All four sides of the 'box' are grooved to accept the bottom. This will cause the bottom to 'float' and will allow for any future expansion of the material. Most importantly, no glue is used at these joints.

Hanging the Drawers. There are, of course, many ways to hang drawers in cabinets. The most popular is to use manufactured *drawer-guide assemblies*. Simply mount this hardware to the drawer and carcase sides. Careful measuring is required; a typical installation is shown on pages 152 and 153.

An alternative to using drawer glides is to build a framework within the carcase for the drawer to slide on. This framework is typical frame-and-panel joinery but because it is used in the interior of the cabinet, it is sometimes referred to as *web framing* (page 156). The rails and stiles add rigidity to the carcase while the center rail acts as a base for the drawer guide. Dust panels are recommended to protect stacked drawers from dust. A notch cut in the back of the drawer or a dadoed track attached to the drawer bottom fits and slides over a hardwood center guide attached to the web frame. Typical web-frame construction is shown on pages 154 and 155.

3 Using a jig, as shown, cut a ¼" wide x ⅜" deep groove on each end of the drawer front, creating two tenons which measure at least ¼ inch thick.

4 Cut the two tenons on the inside surface of the drawer front to the needed length. The resulting tenons should measure no less than ¼ inch long.

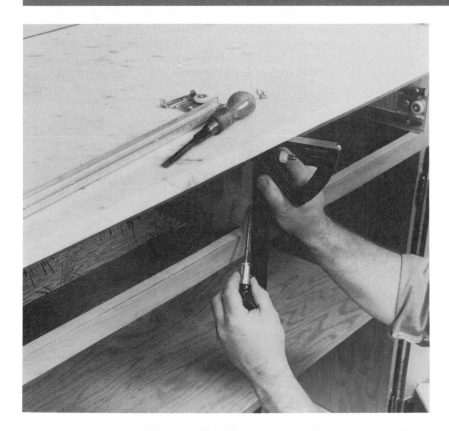

1 Use a combination square to measure the location for the outer glide. Place the handle of the square on the inside top of the carcase and the ruler against the side. With an awl, mark the vertical position of the outer drawer glides according to the height of the drawer and the manufacturer's instructions. Make awl marks in several places the same distance from the top. With a straightedge, scratch a line connecting the points. Measure for the location of the glide on the other side of the carcase in the same way.

3 Read the manufacturer's directions for mounting the inner glides. These are usually similar to the following instructions. Position a combination square's handle on the bottom edge of the drawer so the ruler runs flush against the side of the drawer. Make awl marks at the needed measurement; make additional marks in several places. Scratch an awl line to connect the marks. Place the inner glide on the drawer so the awl line shows through the center of the screw holes and the glide is flush with the drawer front. Make awl marks for the oblong screw holes, drill pilot holes, and mount the inner glide. Repeat on the other side of the drawer.

2 Check the outer glides to see if they must be installed on specific sides of the carcase. This determined, place the bottom edge of one assembly along the awl mark. The front edge should be flush with the carcase. Pull out the glide extension so all screw holes are exposed. Typically, there will be two oblong holes and several round ones. While holding the assembly in place, make awl marks in the oblong holes. Repeat on the other side of the carcase, drill pilot holes, and mount the outer glides.

4 With all hardware temporarily mounted, check the alignment by sliding the drawer closed. When the inner glides hit the stops, tilt the drawer up slightly to slide it all the way into position. Check the alignment and how it fits in the opening. Make necessary adjustments.

5 If the glide assemblies are misaligned, adjust them in one, or both, of the following ways: If the drawer does not close properly, loosen the outer glide screws. Move the glide back or forth as necessary and retighten the screws. If the drawer is not precisely aligned, loosen the inner glide screws and move the glide up or down for

adequate clearance. Retighten the screws; put the drawer back into the opening to recheck the alignment. When the drawer fits correctly, remove it and mark the centers of all circular screw holes on both the inner and outer glides. Drill pilot holes and install the remaining screws to secure the glides.

Now it is time to install the web frame within the carcase. Slide the web frame into dadoes cut in the sides of the carcase. For extra stability, wood screws are used at the front corners of the stiles.

Building the Doors

Up until now, you've been working with a 'box' — sturdy but not too attractive. However, the moment you add the door, your box turns into a beautiful and useful cabinet. The door can be a plain slab, a slab with decorative touches, or a door with elaborate frame-and-panel joinery.

The simplest door you can build is the *slab type*. It is usually made with veneered plywood that has an attractive grain pattern. The edges and ends of the plywood are hidden with veneer tape. You can also glue or dowel wood strips together to form a slab door. The edges of a slab door can be beveled to

TYPICAL WEB FRAME CONSTRUCTION

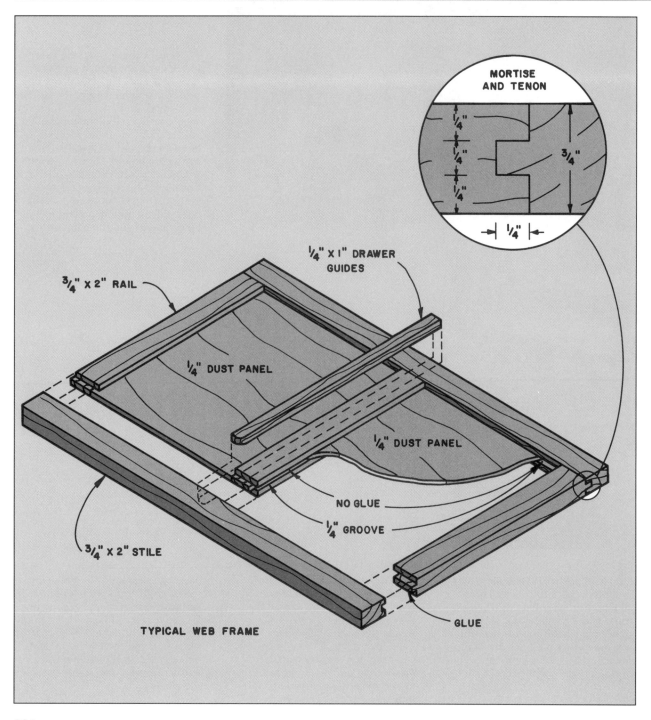

MORTISE AND TENON

¼"
¼"
¼"
¾"
¼"

¼" X 1" DRAWER GUIDES

¾" X 2" RAIL

¼" DUST PANEL

¼" DUST PANEL

NO GLUE

¼" GROOVE

¾" X 2" STILE

TYPICAL WEB FRAME

GLUE

BASIC CABINET CONSTRUCTION

resemble a panel door, or the edges and surface can be routed to give the door a more decorative look.

Frame-and-panel doors are considered by many people to be the most attractive type. Though they may look very difficult to build, a closer look will reveal they are merely grooved stiles and rails surrounding a thinner panel. The stiles and rails are joined with dowels or mortise and tenon joints. Both softwoods and hardwoods can be used for frame-and-panel type of door construction.

Your decor and needs will determine which type of door to use for your cabinet. Whichever one you choose, you'll be able to build it yourself by following the information on pages 157 and 158.

After the door is built, you'll need to select and mount the hinges. Hinges are the hardware that control the door's motion. Several types of hinges are shown earlier in the text on page 34. The process for mounting the most basic type, a butt hinge, is detailed on page 159.

1 Cut a ¼-inch groove on the inside edge of the rails and stiles, and on both edges of the center rail. The dust panels will fit in the grooves. If you're not installing the dust panels, you only need to cut the grooves to fit the tongues.

2 Using the same setup, cut mating tongues in the ends of the rails. A tenoning jig helps hold the stock.

CAUTION Always wear proper eye protection when using power tools (pages 3 and 4). Also, be sure to keep your hands out of the danger zone (page 5).

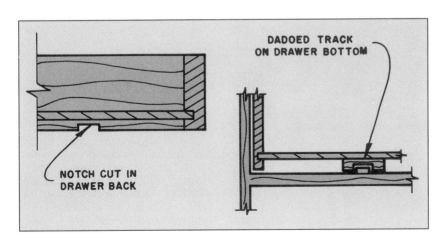

DADOED TRACK ON DRAWER BOTTOM

NOTCH CUT IN DRAWER BACK

3 Make the drawer guide of hardwood and attach it with glue to the center rail. There are several ways to 'notch' the drawer bottom to accommodate the center rail. Shown is a notch cut in the drawer bottom and a dadoed track attached to the bottom of the drawer.

Adding Handles and Pulls

After you've applied the final finish to the cabinet and allowed it to dry completely, add the appropriate hardware to open the drawers and doors. Home improvement centers, hardware stores and catalogs are excellent sources of handles and pulls. When you're installing a drawer pull, you'll get best results if you center it horizontally on the same plane as the drawer's sliding mechanism. When a drawer is extra wide or it's load will be heavy, you should use a pair of pulls, each located approximately 2 to 3 inches from the drawer's sides, carefully aligned.

Door handles are usually positioned vertically. They are installed by drilling a hole for the insertion of a mounting bolt. You should use a drill bit the same size as the bolt and drill entirely through the door; finish by simply adding a washer inside and screwing the handle onto the bolt.

Once the drawer pulls and door handles are installed, it's time to install the cabinets. Installation instructions are found in the next chapter.

Built-In Bookcases

Nothing complements the furniture in a room or decorates a wall more than built-in bookcases. They provide space to hold books, display family photographs, and show off antiques and collectibles. Basic bookcase construction is very similar to that used for case type carcase construction (plywood sides, top and shelves). You can go beyond the basic bookcase by incorporating a pull-down writing desk and a storage area with doors if you desire to create a more spacious wall unit. A good choice of materials to use for built-in bookcases and wall units is hardwood veneered plywood for the sides and shelves, matching wood for the trim, hardboard for the back, and hardware to support the shelves. Typical bookcase construction is shown on pages 160-163.

Touching Up Problem Spots

It's very easy, especially if you're a beginner, to mistakenly 'bruise' your stock in the process of working with it. By clamping wood too tight, nailing incorrectly, or slipping with a tool, you can create a dent, crack, split, or gouge.

If this happens to you, rather than despairingly tossing out costly materials and starting over, you should try to repair the problem area with one of the tips on pages 164-166. Once all mistakes have been mended, you can select and apply the finish of your choice.

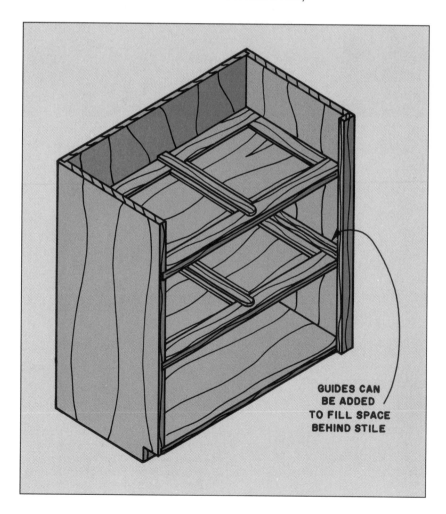

GUIDES CAN
BE ADDED
TO FILL SPACE
BEHIND STILE

Web framing is an alternative to using drawer-glide assemblies to support and guide the drawer. Web framing can be used with case (shown here), frame or panel construction. However, with frame or panel construction, wooden cleats must be mounted to the carcase sides to support the web framing.

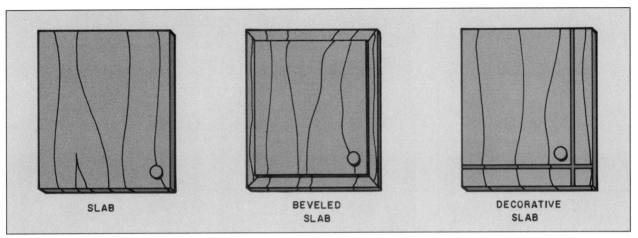

SLAB

BEVELED
SLAB

DECORATIVE
SLAB

Slab. Use glued-up wood or plywood to build the slab door. The slab can be left plain, or if you use glued-up wood you can add decorative touches such as beveled edges and routed grooves. Cover the ends and edges of plywood with veneer tape.

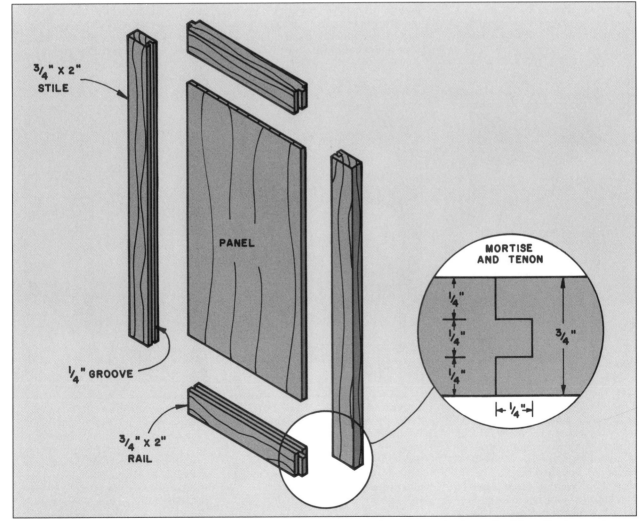

3/4" X 2"
STILE

PANEL

1/4" GROOVE

3/4" X 2"
RAIL

MORTISE
AND TENON

1/4"

1/4"

1/4"

1/4"

3/4"

1/4"

Frame-and-Panel. Cut a 1/4-inch groove on the inside edge of the rails and stiles. Use a table saw, or clamp the pieces to a workbench and use a router to cut the groove. Using the same dado setup, cut mating tongues in the ends of the rails. A tenoning jig helps hold the stock (Step 3 of "Making a Lock Joint" on page 151).

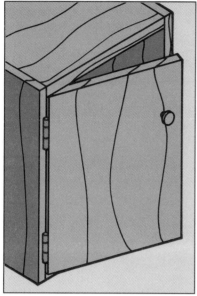

Flush. The flush door fits inside the carcase opening or face frame. It must be cut precisely and hinged carefully so that it will work smoothly.

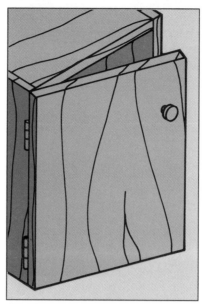

Overlapped. This is the easiest door to cut and install since it covers the entire carcase front. Small imperfections are not critical since the door does not fit inside anything.

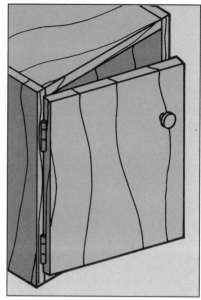

Partially Overlapped. This door is often used on carcases that have face frames. It covers the front opening but does not conceal the entire front of the carcase. Special hinges such as the semi-concealed type are used to mount it.

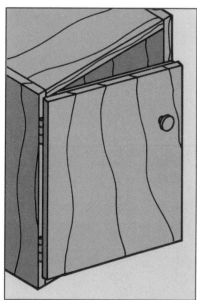

Lipped. The lipped door has a tight seal because it fits inside and over the front of the carcase. The lip is usually created by cutting a rabbet around the inside edges of the door, but another method may also be used. Two differently sized panels may be glued together to form the lip. When measuring for this type of design, remember to allow space for the hinges.

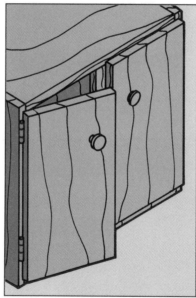

Separated Double. These two doors are separated by a stile. Because of the way they are mounted, the doors operate independently of each other. Although partially overlapped doors are shown, other types may be used.

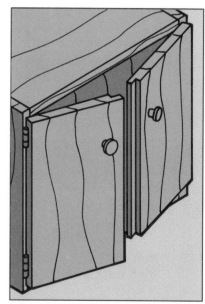

Contiguous Double. When the carcase opening is wider than it is high, the weight of the doors puts a strain on the hinges. A way to remedy this problem is to install contiguous doors. They may be flush, lipped, or overlapped, as shown here. The center edges can be rabbeted so that they fit into each other, as shown, or the edges can be flat to just clear each other when the doors are opened and closed.

Finale — Finishing Your Built-Ins

In truth, to cover this subject thoroughly, it would require a book of its own. Instead, here you will learn enough to understand the steps involved in finishing and the basic kinds of products available. There are actually two main steps involved in the process — *preparing* for the finish and *applying* the finish. The materials for these steps will be discussed separately.

Fillers, Sealers and Abrasives. Holes, cracks and hammer marks all need to be filled to attain a surface that is even and attractive. The most popular substance for this is called *wood putty*. It's available as a powder, and is mixed with water to the desired con-

sistency. This water-based product can be sanded and it responds to stain and paint almost like real wood. Unlike wood dough and similar products, it will not shrink or loosen with time.

If the piece is to be painted, wood putty alone will be fine but if you are using a clear finish, you might also use a *putty stick* for coloring the putty to match the wood. Also, wood putty works well for oil-type finishes, but if the piece will receive a built-up finish such as varnish or polyurethane, you should instead use a paste wood filler. To be sure of the results, especially if you are applying a stain, first test a scrap piece of wood with wood putty or paste wood filler used in combination with a putty stick.

INSTALLING A BUTT HINGE ON A FLUSH DOOR

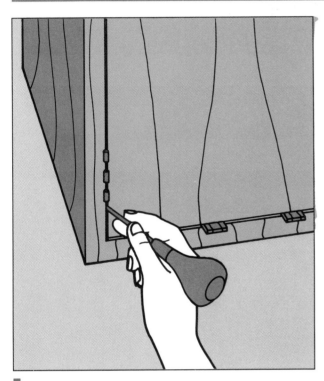

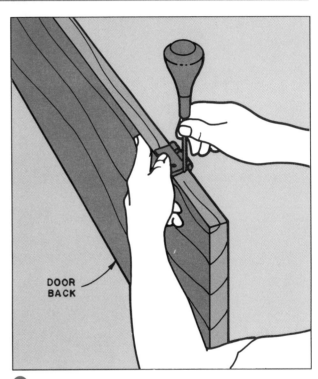

DOOR BACK

1 Remove the pin from the hinge, if possible, and work with separated leaves. Place one leaf on the back of the door with the pin barrel extending just beyond the door's front edge. With an awl, trace an outline of the leaf. With a sharp tool score the marks; also score a line on the edge of the door for the depth of the hinge leaf. Cut a mortise using a hand chisel to make the hinge leaf flush with the surface of the door. Shave the mortise smoothly. Lay the hinge leaf in the mortise so only the barrel extends beyond the door's edge. Mark the screw holes; remove the hinge and drill pilot holes. Attach the leaf. Cut mortises and attach the remaining hinge leaves.

2 Set the cabinet upright and close the door. Use thin slivers of scrap wood at the top and bottom to hold a flush door in place. With an awl, mark the front of the cabinet at the ends of each attached hinge barrel, as shown. Remove the door; then, using the marks for measuring, position the other hinge leaf just inside the cabinet. The pin barrels should extend just beyond the front edge. Trace these leaves, cut mortises, and attach the hinge leaves. (If the hinge pins are not removable, use a combination square to transfer the dimensions of the leaves to the cabinet edge.)

Now that you've seen how to build the individual cabinet components, it's time to plan your cabinets. First decide what type of carcase (case, frame or panel) and shelves (fixed or adjustable) you want; the type of drawers (flush, lipped or overlapped) and how to hang them (draw-glides or web-frames). Choose the style of door (slab or frame-and-panel). Also decide the method you'll use to hang the doors (flush, overlapped, etc.)

To determine the dimensions for your cabinets, measure the space where the cabinets will be installed and then follow the "Standard Dimensions for Cabinets" on page 144.

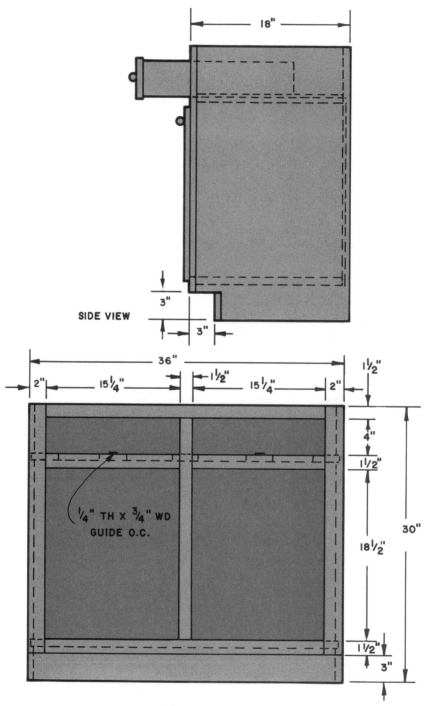

SIDE VIEW

FRONT VIEW

BASIC CABINET CONSTRUCTION

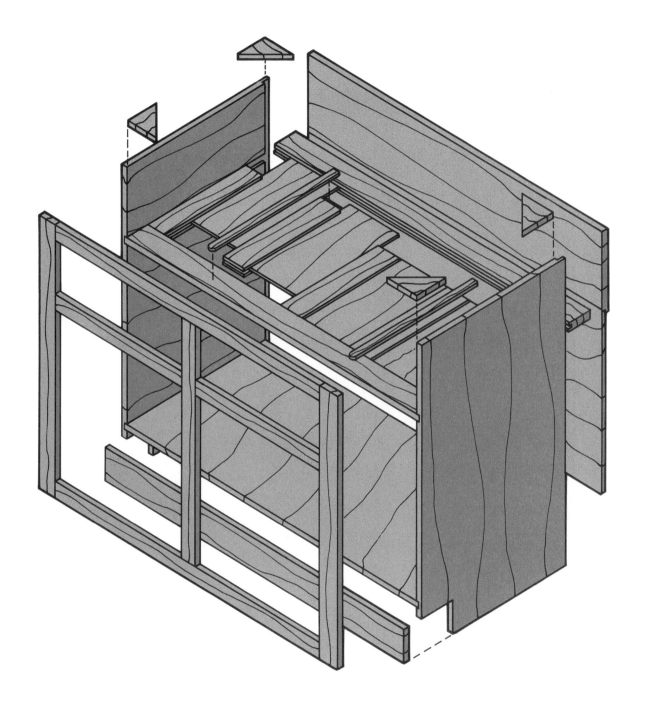

EXPLODED VIEW

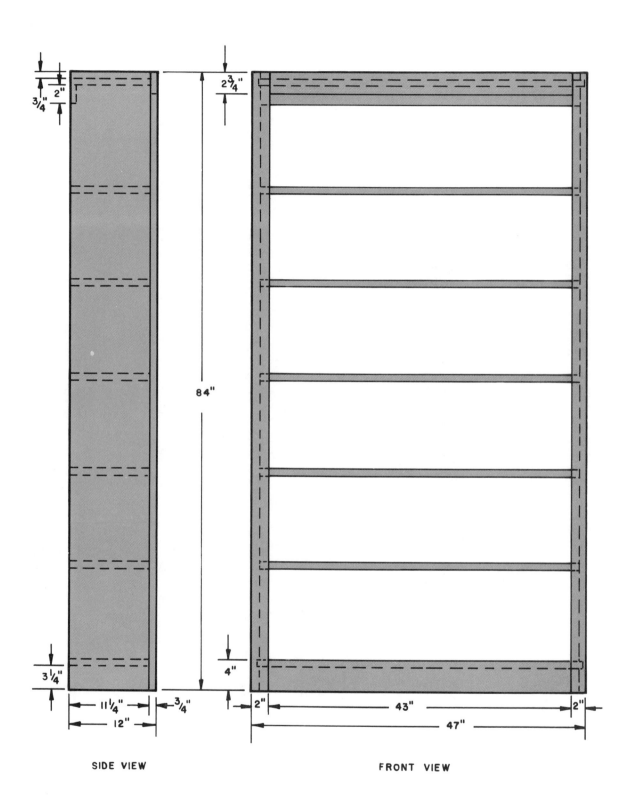

SIDE VIEW   FRONT VIEW

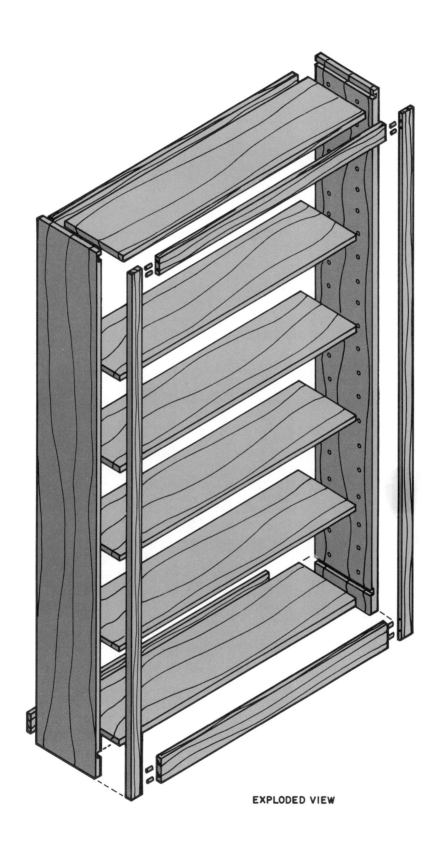

EXPLODED VIEW

1 Prick the dented surface with a pin and apply a few drops of water. Allow it to soak in for a few minutes.

2 Place a piece of aluminum foil over the dent and lay a cloth over the foil. Apply a hot iron to the cloth for a few minutes.

3 Remove the iron, cloth, and foil. Check to see if the dent is raised. For stubborn dents or ones in very hard woods, repeat the process several times.

1 If possible, make this repair to unfinished wood. Use stick shellac which comes in a variety of colors. To a scrap piece of wood, apply the intended final finish; match your shellac to this. Slice off a piece of stick shellac with an old knife which has been heated by a torch flame.

2 Smear the shellac around, forcing it into the crack. When the crack is filled, smooth over the surface of the shellac with the hot knife. Reheat the knife if necessary.

3 After the shellac cools, remove the excess by sanding. The crack should be barely visible as long as it matches the color of the wood.

Sealers are substances for sealing in the stain and filling spots to help prepare for final finishing. A common product for this step is shellac although 'sanding sealers' are also available. An alternative is to choose a stain that stains and seals at the same time, allowing you to skip the sealing step.

Abrasives include sandpaper and steel wool of various grades. For rough sanding, use medium (80#) sandpaper; for general smoothing, use fine grade (120#) and for the final stage, use very fine (180# or 220#). Sandpapers are available in several different types of materials; durable aluminum oxide and less expensive but not as long-lasting garnet and flint. After sanding a surface, use the finer grades of steel wool, #1 to #3/0 for an even smoother surface. A light rubbing with fine steel wool will smooth the surface after a stain has been applied.

Choosing a Finish. There are many, many types of finishes available. In choosing the appropriate one for your project you should consider these

REPAIRING A SMALL GOUGE

1 Cut a plug from the same stock that was used to build the project, or use a piece of stock that closely matches the color and grain pattern of the gouged wood. Make sure that the plug is just a little thicker than the depth of the hole.

2 Drill out the gouged area. Position the plug so the grain runs in the same direction as the surrounding area, then glue the plug in place.

3 After the glue has dried, sand the plug flush. If the grain of the plug runs in the same direction as the surrounding area, your repair work will not be very noticeable.

CHOOSING THE APPROPRIATE FINISH

FINISH	CHARACTERISTICS	TYPES AVAILABLE	ADVANTAGES	DISADVANTAGES
Penetrating Oil Stains	Create a lustre and a thin film with several applications.	Linseed oil, tung oil, Danish oil, teak oil, salad-bowl oil, mineral oil, beeswax with turpentine.	Easy to apply.	Slow-drying, not resistant to alcohol or water.
Shellac	Creates a hard, transparent film. Colorless or amber.	Natural and white. Natural is orange in color. White is almost color-less. Thinned by adding alcohol	When thinned, easy to use. Dries fast; easy to sand. Good for sealing wood.	When thick, difficult to apply. Many applications needed for durability.
Polyurethane Varnish	Creates a hard, transparent film that slightly darkens wood.	Available in flat, glossy, and semi-gloss.	Extremely resistant to wear and spills. Easy to apply.	Difficult to sand; darkens with age. Not recommended over shellac.
Acrylic Varnish	Creates a hard, crystal-clear film.	Available in flat, glossy, and semi-gloss.	Shows off true color of wood. Will not yellow with age.	Difficult to apply when brushed on. Limited resistance to wear and spills.
Alkyd Varnish	Creates a hard, transparent film that slightly darkens wood.	Available in flat, glossy, and semi-gloss.	Moderately resistant to wear and spills. Easy to apply.	Not for heavily-used projects.
Epoxy Varnish	Creates a hard, transparent film that slightly darkens wood.	Available in flat, glossy, and semi-gloss.	Extremely resistant to wear and spills. May also be used on metal.	Difficult to sand; darkens with age. Manufacturer's instructions must be followed exactly.
Lacquer	Creates a hard, thin, transparent film.	Clear, plus a wide range of colors.	Fast drying. Excellent resistance to wear and spills.	Difficult to apply with a brush. Not recommended over other finishes.
Alkyd Enamel	Creates a hard opaque surface.	Available in flat, glossy, and semi-gloss. Wide range of colors.	Easy to apply. Moderately resistant to wear and spills.	Not for heavily used projects.
Latex Enamel	Creates a firm opaque surface.	Available in flat, glossy, and semi-gloss. Wide range of colors.	Easy to apply. Easy to clean up.	Not for heavily used projects. Surface must be sealed first with shellac or an enamel undercoat.

issues: How will the project be used? If it will receive hard use, you should give it a tough finish. How do you want it to look? If you want to change the color of the wood, make it look glossy, or cover it entirely, all these can be done with the finish. What species of wood are you covering? The type of wood you're working with might affect your choice of finish.

The chart on page 167 lists several types of finishes. This chart will give you a general idea of finishes in regard to their appearance, their advantages, and their disadvantages. However, the only way to know exactly what your results will be is to do a little research on your own. Ask friends who are woodworkers which finishes they have used. Look for displays in hardware and paint stores. Within the boundaries of what your taste will permit, you should choose a product that offers ease of use, especially if you're a beginner.

Stains are most often used to color wood and hide blemishes before the final finish is applied. There are several types available: *water stains, pigmented oil stains,* and *penetrating oil stains.* Penetrating oil stains will usually yield the most pleasing results. A chief consideration in choosing the stain is the type of material you're staining. For example, pigmented oil stains do not cover hardwoods well. Also, particleboard will not accept a water stain well. Choose the best product for your wood and then apply it according to the manufacturer's directions. It's always best to use a 'test board' first to see if you're going to be

TOOLS FOR APPLYING A FINISH

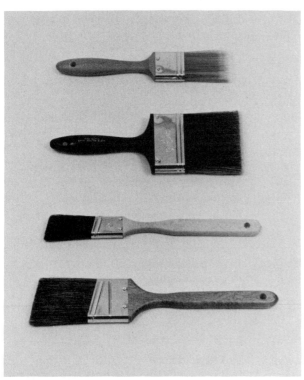

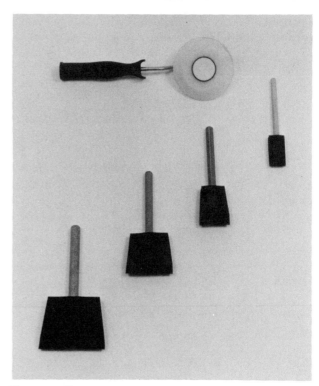

Brushes come in many different sizes and shapes, and with natural or synthetic bristles. Choose brushes with natural or polyester bristles to apply oil-based finishes; and ones with polyester, polypropylene or nylon bristles to apply water-based finishes. The size and shape of the brush you choose are determined by the size and shape of your project. For instance, use a 3-inch wide brush with a chiseled edge for doors and shelves and a 1- or 2-inch beveled or chiseled edge for trim and molding. When choosing brushes, remember that the quality of the brush dictates the speed of application and the appearance of the finished surface.

Foam applicators, available in many sizes and several shapes, are used for all finishes except lacquer and shellac. They are excellent for applying polyurethane. They can be cleaned and reused or, since they are inexpensive, thrown away.

BASIC CABINET CONSTRUCTION

satisfied with the stain. Most penetrating oil stains can be tinted to a different color by adding common artist's pigments.

Application Tools. Finishes are applied by a number of methods. You can use a brush, rag, foam-type applicator, or spray gun (pages 168 and 169). It's best to choose a quality applicator, especially if you're using brushes.

Choose the types of brushes and brush sizes you need according to your project and the type of finish you're using. Varnish brushes, angled and oval sash brushes, even artist's brushes or wall brushes may be used. Brushes must be well taken care of to preserve their life. This involves carefully cleaning them after each use following the steps on page 170.

Application Tips. No matter what kind of finish you use, application should be accomplished with care and precision. Use the following checklist as a general guide:

■ Read *every* label. Each product you use has its own features and application techniques, so following directions is the only way to be sure of success. Pay special attention to the WARNINGS on containers.

■ Start with a clean, dust-free surface; always use a tack cloth to remove all dust and residue.

■ Work in a well-ventilated area.

■ Try to work at eye level, and use adequate lighting.

■ Make test applications to see how the finish will change the color of the wood.

■ Do the difficult parts first.

A handmade applicator — an ordinary sponge covered with a soft rag — can be used to apply penetrating oil stains.

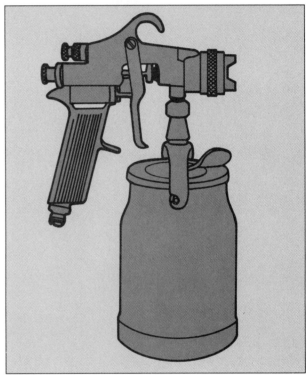

Spray equipment can save time. Depending on the size and scope of your project, you might want to buy or rent spray equipment to apply your finishes. Get plenty of practice with a spray gun before you finish good materials, follow the manufacturer's instructions and always work with safety gear in the proper environment.

1 When you've finished applying your finish, remove the excess by wiping the brush thoroughly with a clean rag or paper towel.

2 If an *oil-based finish* was used, swish the brush in the proper solvent. Remove the excess solvent with a clean cloth or paper towel, wrap the bristles in paper, and then hang the brush to dry.

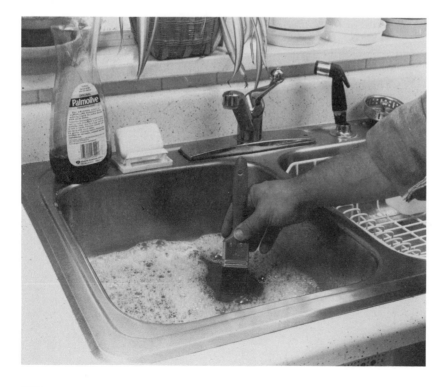

3 If you used a *water-based finish*, use soap and water for cleanup; be sure to rinse thoroughly. Remove excess water with a clean rag or paper towel. Hang the brush to dry.

- Use a pipe cleaner or toothbrush for extremely hard-to-reach spots.
- Extend the drying time for stains and clear finishes when they're applied in cold and humid environments.
- Once you've finished the application, leave the workshop. Walking around stirs up dust — dust that can ruin your finish.

A Finishing Schedule. Each project is different, but these are the typical steps involved in preparing for and applying a finish. Not every step will be used for every project, but if used, they should be done in the following order:

1️⃣ Fill holes, cracks and hammer marks with wood putty or wood dough. This creates a smooth, even surface for the finish.

2️⃣ Smooth the surface with sandpaper or steel wool. Wipe away dust and residue with a tack cloth.

3️⃣ Apply a stain as shown below if a special wood color or grain highlighting is desired.

4️⃣ Apply a sealer if your wood is coarse textured. Sealers allow for even application of a finish. They also prevent certain stains from bleeding into and clouding the finish.

5️⃣ Apply the final finish.

6️⃣ Use fine sandpaper or steel wool between successive coats of finish to smooth the surface.

7️⃣ Apply paste wax and buff the surface after the final coat.

APPLYING A STAIN

1 Apply a wood stain *with* the grain. Begin each stroke a few inches from the last, and work in the direction of the absorbed stain. Using this method, you'll get fewer dark, irregular 'overlapping' marks.

2 After allowing the stain to penetrate in a time period specified by the manufacturer, wipe the surface with a clean cloth. You might need to apply the stain and wipe the project in sections.

10

Installing Cabinets

If you've ever hired a contractor or cabinet dealer to install cabinets in your home, you'll probably remember that the workers spent less than a day doing moderately heavy semi-skilled work. If you watched them you might have been amazed at how easy the process looked.

Installing cabinets is not really difficult. Like so many of the carpentry skills discussed previously in this text, it merely requires careful measuring and constant checks of level and plumb. The goal is two-fold: to *securely* attach the cabinets to walls and to make them *fit* like a glove, or, at least, to make them appear to fit perfectly.

Here you'll learn the proper order of installation. You'll also be provided with tips that cabinet installers use. Whether you build your own cabinets or order them from a supplier, you can use this information to install them in a professional manner.

An integral part of kitchen and bath cabinets is the countertop which is usually made of plastic laminate. You'll see how to laminate a countertop, make cutouts in the surface, plus you'll learn the process of anchoring the top to the carcase. The information is all here, so hang in there...and then hang the cabinets!

Installation —
Making Cabinets Permanent

Your cabinets are built and finished; now it's time to anchor them to the walls and make them a permanent part of the room. Two things will be accomplished. First, the cabinets should be anchored solidly enough to carry the load for which they were built. Second, the finished look should be such that the cabinets become an integral part of the room. Apart from these considerations, you should also make allowances in your cabinet plans to accommodate the wiring and plumbing.

Even if you haven't built your own cabinets, you can use these guidelines to install ready-mades.

General Instructions. When you transport the cabinets to the work site, wrap them with blankets and be very careful to avoid damage — both to the cabinets and to drywall or other wall surfaces. Once the cabinets are at the site, you should have all the needed installation tools on hand to avoid interruptions. The sequence of installation is as follows:

1. Install the base cabinets.
2. Install the wall cabinets.
3. Build and laminate the countertop.
4. Mount the countertop and make cutouts.
5. Add the trim.

The installation of base and wall cabinets involves constant checks of level and plumb and the use of shims. This is even more likely if you're working in an older home; in this instance, check the room

INSTALLING A BASE CABINET

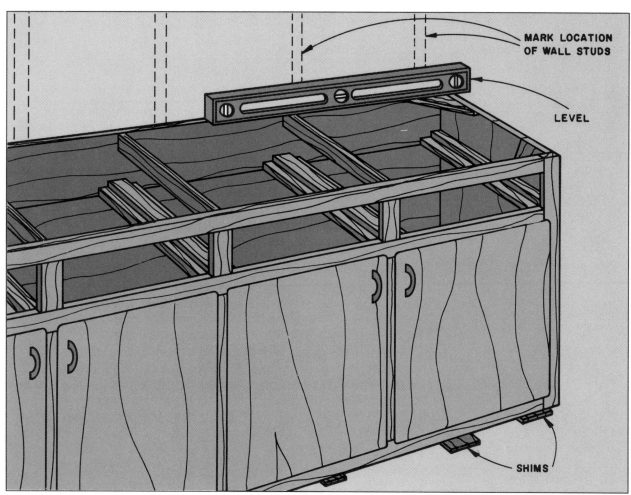

Locate and mark the wall studs. Using typical stud measurements, make marks on the wall in the space between where the base and the wall cabinets will be mounted. Level the cabinet(s) both front to back and horizontally by inserting shims under the cabinet.

Drill pilot holes through the cabinet back and into the stud. Mount the cabinet with #12 round-head wood screws and washers that are long enough to penetrate the cabinet back, the wall covering, and at least 1 inch into the stud. Remove the shims, cut off the part that

extends beyond the cabinet, and reinsert. The shims and the gap will be hidden by a base shoe molding.

you're working in before you begin, so you'll be aware of any major problems. For minor problems at wall surfaces, narrow trim and shoe molding will hide the gaps. A long row of cabinets should match up perfectly at every edge.

Cabinetmakers have their preferences, but generally it's best to install the base cabinets before the wall cabinets. If your plan includes corner cabinets, install these first.

Installing the Base Cabinets

First, locate and mark the wall studs. Use a stud finder or measure out from known studs — a corner has double studs and electrical outlet boxes are

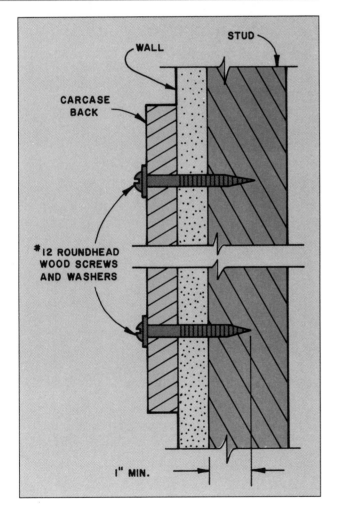

mounted to studs. At least they should be. If all else fails, use a thin nail and simply hammer into the wall until you locate the studs. Do this in an area that will be covered by the cabinets. To install the base cabinets use the instructions on page 174 and this page.

Hanging the Wall Cabinets

Lay the core material you'll be using for the laminated countertop on the top of the base cabinets. This will allow the base cabinets to support the wall cabinets during installation, with the help of 'horses' or wedge boards.

Use either wood screws, hollow-wall fasteners, or toggle bolts to anchor the cabinets to the wall. Each cabinet should be fastened with at least two fasteners at the top and at least two fasteners at the bottom. If a cabinet spans only one stud, hollow-wall fasteners or toggle bolts must be used to install the wall cabinet securely. Instructions for installing wall cabinets can be found on pages 176 and 177.

Tips about Tops.   Two popular choices for countertops are plastic laminate and natural butcher block (page 178). You can purchase pre-made laminate countertops or you can laminate your own. Butcher block can be purchased or constructed of strips of hardwood, usually maple.

The process of applying plastic laminate is not too difficult (pages 179 and 180). Particularly if the top is manageably small, it's simply a matter of using a special adhesive and handling the fragile material carefully. A router with a special laminate trimming bit is needed for the edging process.

Never install a *finished* countertop before you install the wall cabinets. You could easily damage the surface by working above it.

Install the countertop to the base cabinets in a number of ways. Nail it with 6d finishing nails or coated box nails, or fasten it in place by driving screws up through pre-drilled wood cleats on the cabinet sides and back. Be careful not to drive the screws all the way through the core material; this will damage the laminate. Yet another technique for attaching the countertop, if your cabinet design permits, is to use mirror clips (page 181).

With the countertop fastened, next make any cutouts in it for sinks or range tops. Use the instructions on page 181.

Trim — Hiding the Gaps

One way to cover up the gaps where the cabinets meet the wall or floor, or to just add a decorative touch is to use molding. Make the molding from the same material the cabinets are made of, or purchase molding and stain it to match the cabinets. Molding, called flat scribe molding, is readily available to match manufactured cabinets.

Install molding where the cabinet meets the soffit, the wall, and the floor (page 181).

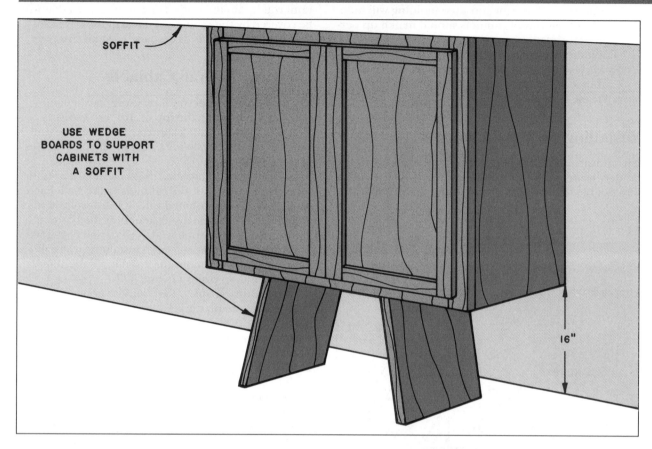

SOFFIT

USE WEDGE
BOARDS TO SUPPORT
CABINETS WITH
A SOFFIT

16"

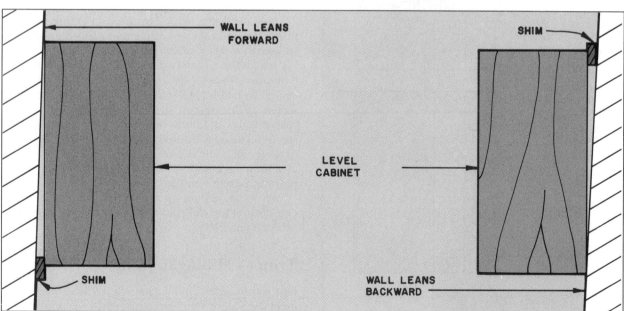

WALL LEANS
FORWARD

SHIM

LEVEL
CABINET

SHIM

WALL LEANS
BACKWARD

Typical height between base and wall cabinets is 16 inches. Use wedge boards made of ½-inch plywood to support cabinets when installing them beneath a soffit. For a 'free-standing' installation where there is no soffit, use small 'horses'.

With a helper, first lift the cabinet onto the base cabinet; then carefully raise it and insert supports. Check for level; add shims behind the cabinet if necessary. Then drill pilot holes through the cabinet back and into the wall studs. Use #12 roundhead wood screws and washers. The screws should be long enough to penetrate the cabinet back, the wall covering, and penetrate the stud by at least 1 inch. If the cabinet is not wide enough to span two wall studs, you'll need to use hollow-wall fasteners or similar hardware to fasten the cabinet securely to the wall.

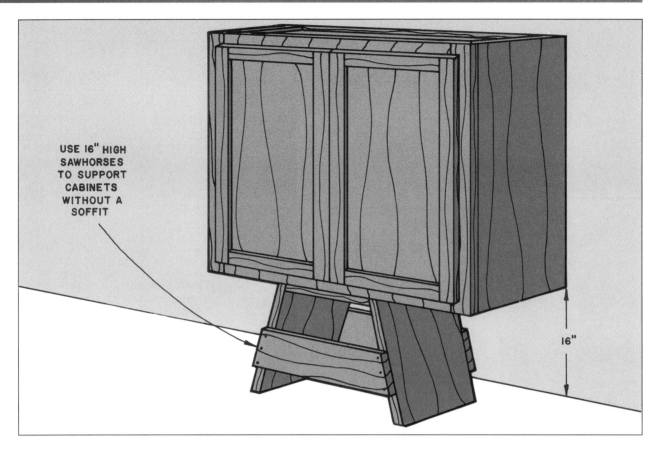

USE 16" HIGH
SAWHORSES
TO SUPPORT
CABINETS
WITHOUT A
SOFFIT

16"

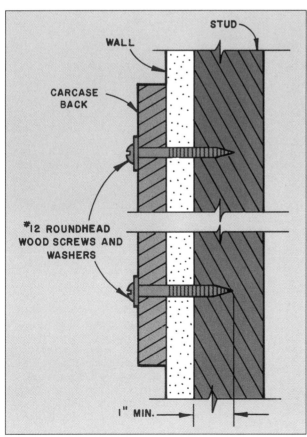

WALL

STUD

CARCASE
BACK

#12 ROUNDHEAD
WOOD SCREWS AND
WASHERS

1" MIN.

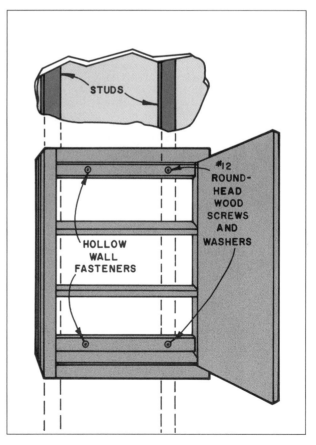

STUDS

#12
ROUND-
HEAD
WOOD
SCREWS
AND
WASHERS

HOLLOW
WALL
FASTENERS

Plastic Laminate. The most common material for countertops, chosen for its durability, is plastic laminate. You can make your own or you can buy custom-made plastic laminate countertops in any size, shape or color you desire. An inexpensive alternative is to purchase pre-made laminated countertops at building supply stores. These usually come in six to twelve foot straight lengths and in several colors.

Butcher Block. The butcher block countertop, made of 'real wood', is functional in that it can be used as a cutting board. You can construct your own butcher block top or purchase one ready-made like the one shown here. Since they are so expensive, most people use only small sections of them or top small cabinets with them.

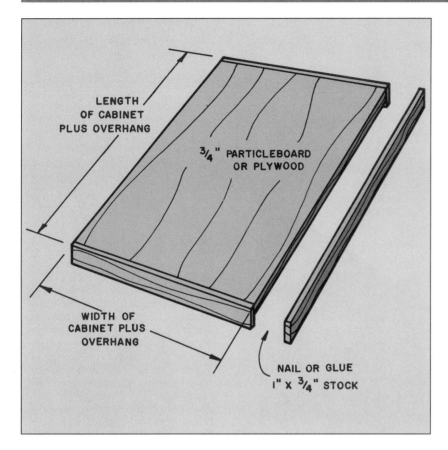

LENGTH
OF CABINET
PLUS OVERHANG

³/₄" PARTICLEBOARD
OR PLYWOOD

WIDTH OF
CABINET PLUS
OVERHANG

NAIL OR GLUE
1" X ³/₄" STOCK

1 Plastic laminates are sold by the square foot, and the standard thickness used for countertops is $\frac{1}{16}$ inch thick. Particleboard is the most common core material. However plywood with no voids in the surface where the laminate will be attached can also be used. Measure and cut the core material to size. Allow for an overhang, typically $1\frac{1}{2}$ inches. Then nail or glue strips of wood around the entire edge to increase the thickness of the countertop to $1\frac{1}{2}$ inches.

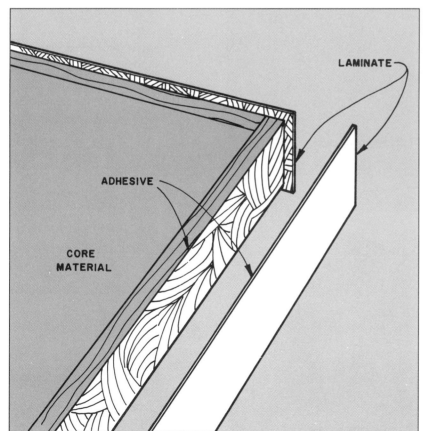

LAMINATE

ADHESIVE

CORE
MATERIAL

2 Cut 2-inch wide strips of laminate for the edges. The extra $\frac{1}{2}$ inch allows for trimming. Apply contact cement to the edges of the core material and edge pieces of laminate. Allow to dry according to the manufacturer's directions. When the surfaces are completely dry, install the edge pieces. Bonding is immediate, so work carefully. Trim off the excess using a router with a laminate trimming bit. Clean up any dust and debris.

INSTALLING CABINETS

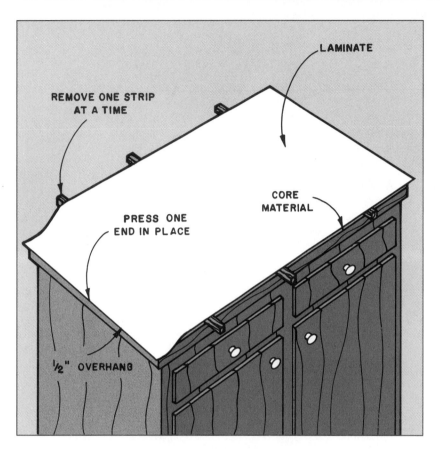

3 Cut the laminate for the top oversize, ½ inch on all four sides. Apply cement to the core material and the laminate. Allow to dry. Before you install the laminate, it's best to lay strips of wood or laminate scraps across the core material so the two surfaces do not touch until they are positioned properly. Once the core and the laminate are in position, slowly remove the strips one at a time starting at one end. Press and smooth the laminate into position. Trim and smooth the edge using a router with a laminate trimming bit.

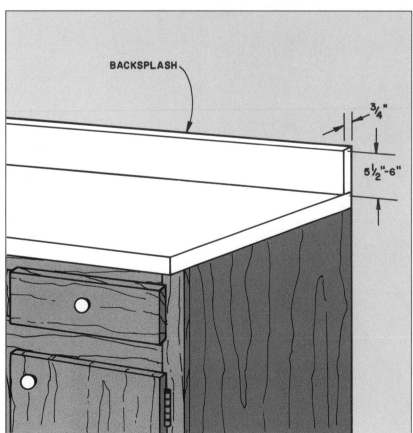

4 Generally, the backsplash is made out of ¾-inch particleboard or plywood covered with plastic laminate. Typical height is 5½ to 6 inches. The backsplash is installed with screws from beneath the countertop and sealed at the seams with high quality caulk.

INSTALLING CABINETS

Attaching the Countertop. Use this easy method to secure a countertop. Into a ⅛-inch groove, slip mirror clips. Screw these to the underside of the countertop.

Making Cutouts in Countertops. To mark for the cutout, use the template provided with the sink or range top, or measure the unit and make a template. Drill a starter hole inside the cutline. Use a sabre saw to make the cutout. Test-fit the unit to make sure it fits without binding. Trim the cutout if necessary.

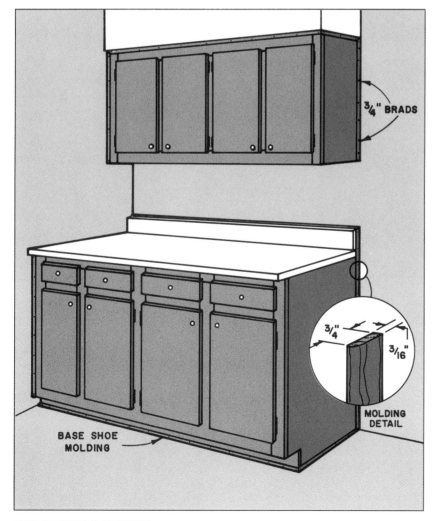

Trim. After the cabinets and countertop are installed, add molding for a decorative touch or to conceal gaps where the cabinets meet the wall, soffit, and floor. Press the thin, flexible strips against the cabinets and secure with ¾-inch brads. Set the nail heads.

Special
Additions

11

Doors
and
Windows

In previous chapters you've seen how openings for standard pre-hung windows and doors are framed and how they are installed in the framing. Now it's time to go one step further. Give your home that special look that will set it apart from all the others in your neighborhood — install custom-made doors and windows.

Your entrance door is a visitor's first closeup look at your home. Why not make it a lasting impression? You can do this with a door that you make yourself or a prefabricated one with your own special touches added. In this chapter there are instructions for making your own doors and ideas for enhancing them with molding and small windows.

Special stationary (non-opening) windows located in strategic spots give your home a unique look. In this chapter you'll learn how to frame and install round, octagonal, semi-circular, and other odd-shaped windows. Instructions for installing drywall and paneling around these windows are also included.

The framing and installation of greenhouse windows, skylights, and bay windows are also covered in this chapter.

Building a Special Door for Your Home

The front door of your home should complement your home's architecture and decor. It's the first thing that visitors to your home will see. So it's undoubtedly important to you and your family that it be special.

You can shop for doors at home improvement centers and building supply stores. There you'll find many different designs and types of doors made of aluminum, plywood and solid wood. They are also available in a wide range of prices. Choose the one that best suits your needs and budget.

If you want a door that features a really unique design or one that you can show off proudly, build your own door. Make your door out of hardwood — cherry, oak or walnut are excellent choices.

There are two basic styles of doors — flush and panel. Most flush doors are simply made of large panels that cover both sides of an internal framework. A panel door consists of panels, stiles, rails and mullions that are joined with dowels or mortise and tenon joints. Solid wood panel doors are the most attractive. However they take considerably more time to build than flush doors but are definitely worth the effort. Making doors requires skill, precise measurements, straight and true stock, exacting joinery and good finishing techniques. If you are installing a window in your door, you must use tempered glass or rigid plastic for safety.

Use the many door designs on page 187 to help design your door and follow the illustrations and instructions below to build a panel door.

The opening for your door is framed as shown on page 86. In order to hang a door in the opening, you will need to construct and install a door frame assembly (page 188).

When your door is completed and the frame assembly installed, hang the door according to the instructions on page 117. Then apply the finish of your choice and install the hardware. Hardware stores or catalogs are great sources of standard and unusual hardware.

BUILDING A PANEL DOOR

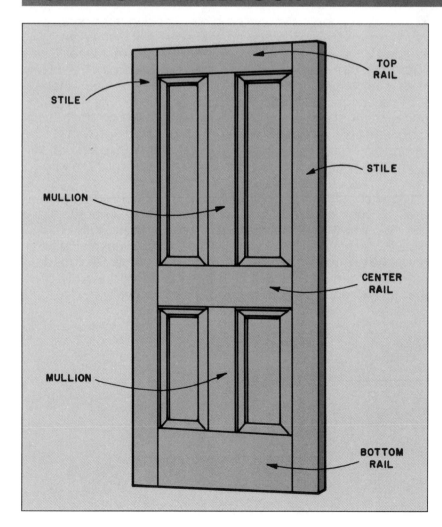

STILE

TOP RAIL

STILE

MULLION

CENTER RAIL

MULLION

BOTTOM RAIL

Panel doors are a framework of stiles (vertical members), rails (horizontal members), and mullions (intermediate vertical members) with panels that fit within it. Use hardwood, preferably oak, cherry or walnut, that is free of knots and blemishes for all pieces.

Cut the stiles, rails and mullions to size. Cut grooves in the edges of each to accommodate the panels. Then drill the holes for the dowels (page 49) or make mortise and tenon joints (page 51) on the ends. Unless you have access to wide lumber, it will be necessary to glue up stock for the panels. For a less bulky look, use a planer to decrease the thickness of the panels. Cut chamfers on the panels. Rout mortises for the hinges and drill holes for the other hardware.

Assemble the door components with waterproof glue and apply the finish of your choice. Where the finish is applied to a door makes a difference in weatherproofing. Finishing the top and bottom edges of a door is every bit as important as finishing the visible surfaces. When the finish dries, mount the hardware. Hang the door according to the instructions on page 117.

FLUSH DOORS

PANEL DOORS

Designing Your Door. Choose a door design that will match the architecture of your home. Basically, there are two styles of exterior doors — flush and panel. Designs for several flush and panel doors, some with windows, are shown here.

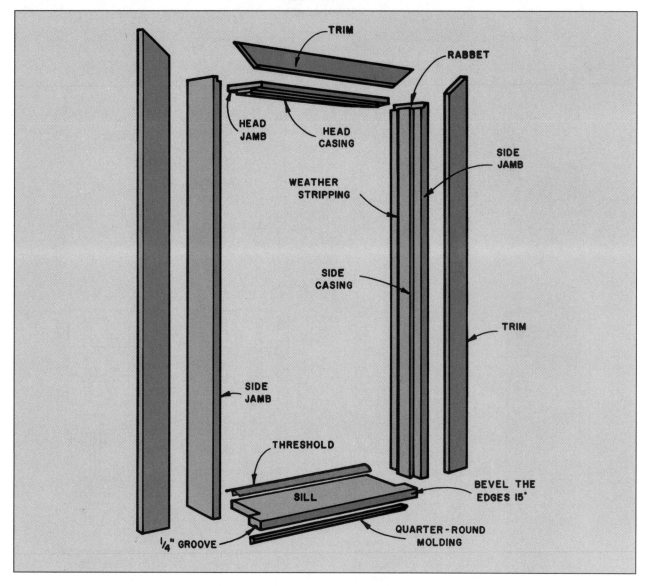

The main components of the door frame are the sill, doorjamb, casing, threshold and trim. If you are an experienced woodworker, you can build the door frame. However, because of the precision required, it is highly recommended that you purchase the components of the door frame at a home improvement center or building supply store.

The *sill* is the lower member of the door frame. It is made of hardwood, usually oak, for wear resistance. If a soft wood is used, you must add wear strips and a metal nosing. The sill must have a bevel on the front and back edges and a ¼-inch groove on the underside near the front to allow water to run off easily.

The *doorjamb* is the part of the door frame that fits inside the rough framing of the door opening. The jamb has three parts: two side jambs and a head jamb across the top. A rabbet, the width of the material used for the jambs, is cut in the upper end of the side jambs.

The *casing* consists of two side casings and a head casing. It is attached to the doorjamb to form a stop. The stop is the part of the doorjamb which the face of the door closes against.

Assemble the doorjamb and the sill. Before installing this assembly, line the rough opening with building paper and attach flashing to the top of the head jamb.

After setting the assembly in the opening, check it for plumb. Add shims between the trimmer studs and the side jambs where needed. Check

the sill for level, and shim it as needed. Position the casing and nail through the casing, jamb, shims and partway into the trimmer studs and header. Drive all the nails in partway. Make a final check of level and plumb, then finish driving the nails. Set all nails using a nail set.

After the finish flooring is in place, install a hardwood or metal *threshold* over the gap between the flooring and the sill.

Cut and add an exterior trim around the door frame. Use a nail set to set the nail heads, and caulk where the trim meets the door frame and the siding. Nail a piece of quarter-round molding to the underside of the sill and install weatherstripping along the inside edge of the head and side casings.

Framing and Installing Special Windows

Special windows may look complicated to frame but the framing process is basically the same as for standard windows with a few exceptions. In some instances the framing is much easier because most of these windows are stationary.

You can make or buy these special windows. Unless you are an experienced woodworker with a well-equipped shop, it is recommended that you purchase the windows. You can frame and install them yourself, then add decorative trim and molding.

Round and Octagonal Windows

Usually round and octagonal windows are small and stationary. They can be located in a stairwell, hallway, bathroom or in the eave of a vaulted ceiling. They will let in a substantial amount of light while adding an interesting effect to the room as well as to the exterior of the house.

Instructions for framing and installing round and octagonal windows are found below and continue through page 196. You can also frame and install small hexagonal, diamond, and triangle shaped windows by altering the instructions slightly.

FRAMING AND INSTALLING A ROUND WINDOW

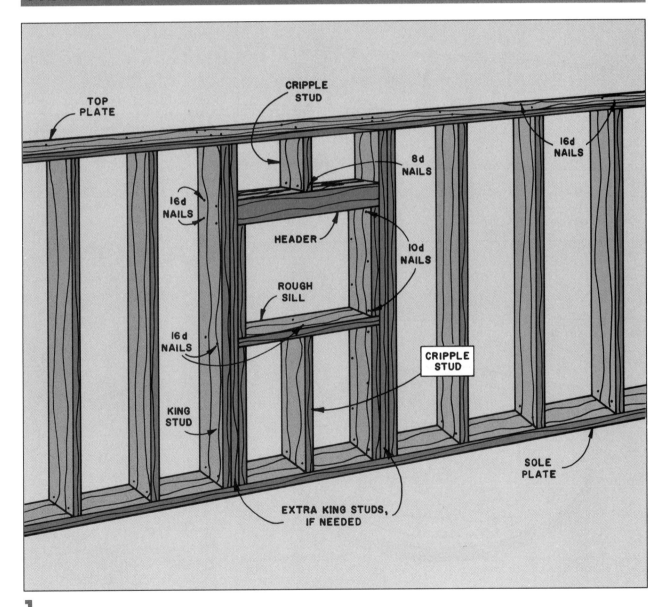

1 During the framing of the structure, frame the opening for the window according to the instructions on page 87. If needed, nail extra king studs on one or both sides of the opening as shown to accommodate the window.

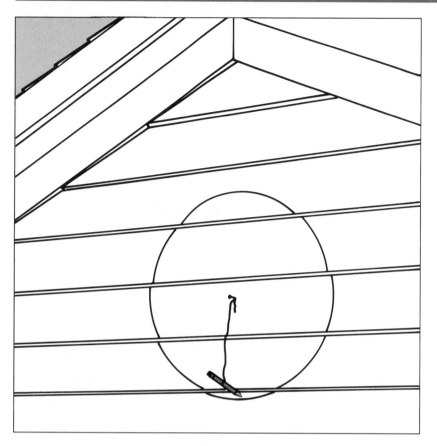

2 Mark the center of the window on the inside of the sheathing. At the mark, drill a small diameter hole through the sheathing and the siding. On the outside of the structure, drive a nail into the drilled hole. Attach a pencil to a piece of string and tie it to the nail. The length of the string between the pencil and the nail should be the radius of the window unit. Draw the circle, as shown, and use a reciprocating or sabre saw to cut out the opening. Remove any siding that is attached to the cutout. Save the cutout for later use.

CAUTION

Always wear proper eye protection when using power tools (pages 3 and 4). Also, be sure to keep your hands out of the danger zone (page 5).

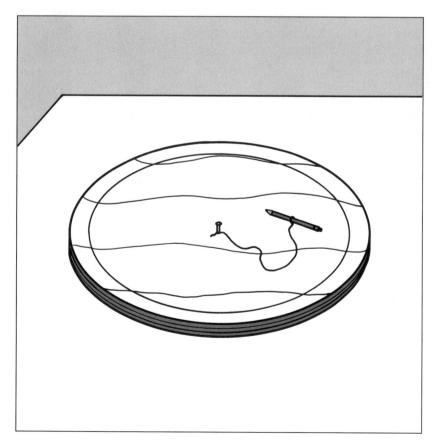

4 Draw a circle on the cutout. The diameter of the circle is equal to the outside diameter of the window frame. Cut out the ring using a sabre saw, scroll saw, jigsaw, or bandsaw. On the outside of the structure, nail the ring to the edges of the braces and trimmer studs. Caulk the space between the ring and the sheathing.

DOORS AND WINDOWS

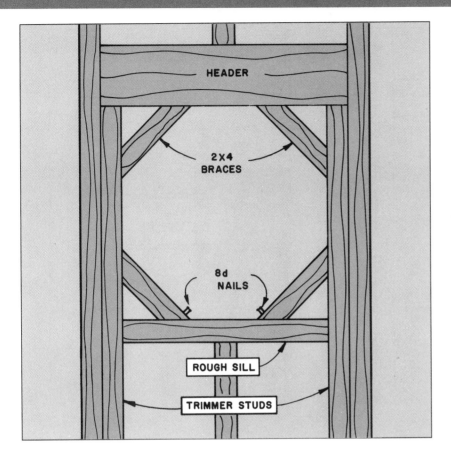

3 Nail 2 x 4 braces to the framing members. To determine the length of the braces, multiply the diameter of the opening by 0.4. Cut the ends of the braces at 45° angles. Toenail the braces to the header, trimmer studs, and rough sill.

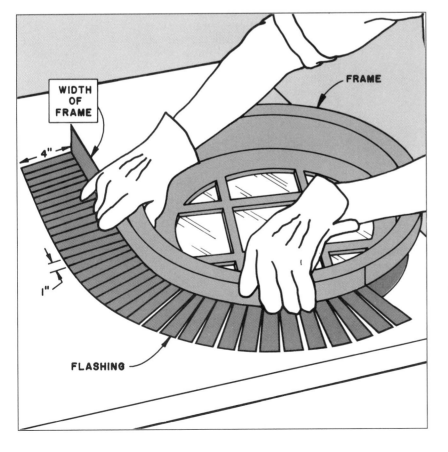

5 Cut a piece of flashing the length of one half the circumference of the window frame and 4 inches wider than its thickness. Cut 4-inch long by 1-inch wide tabs in the flashing. Bend the tabs at right angles, then form the flashing by wrapping it around the frame. From outside the structure, insert the tabs of the flashing between the sheathing and the siding at the top of the opening.

DOORS AND WINDOWS

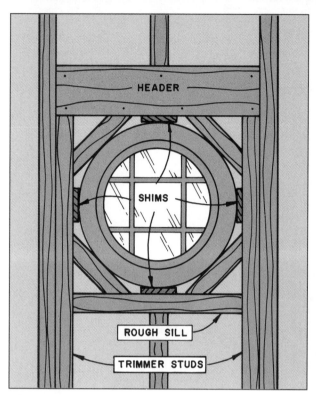

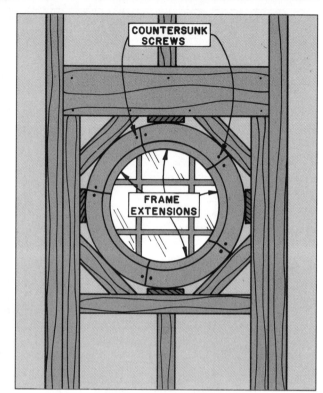

6 Have a helper hold the window in the opening from the outside while you drive shims between the window frame and the framing members in the following sequence: rough sill, header and trimmer studs. Have your helper nail the frame to the sheathing and framing members. Then caulk above the flashing and where the frame meets the siding.

7 Cut off the ends of the shims. Glue the frame extensions, that came with the window, to the frame and anchor them with countersunk screws. Using a block plane, trim the frame extensions flush with the framing members. Follow the instructions on page 202 to make the cutouts in the drywall or paneling for the round window.

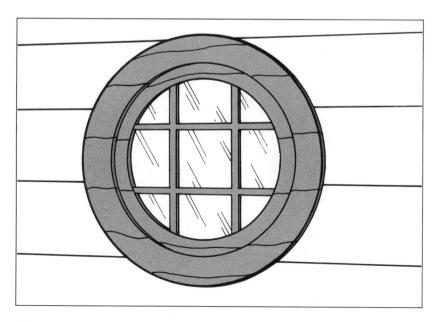

8 If the window did not come with interior and exterior trim, or if you want trim that matches or complements the other trim on the structure, circular trim needs to be made. Unless you are an experienced woodworker with a well-equipped workshop, have a cabinetmaker make the circular trim for you. You can install it yourself.

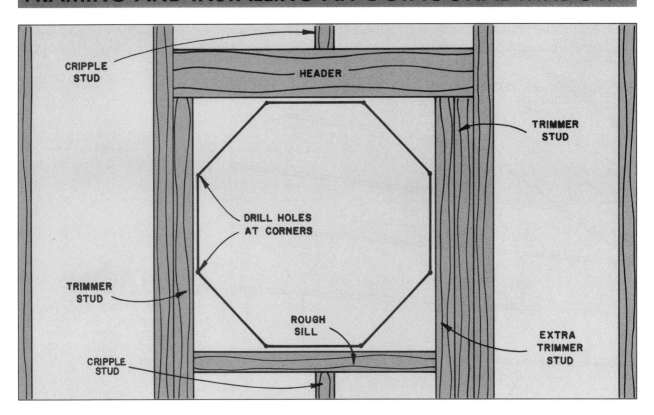

CRIPPLE STUD

HEADER

TRIMMER STUD

DRILL HOLES AT CORNERS

TRIMMER STUD

ROUGH SILL

EXTRA TRIMMER STUD

CRIPPLE STUD

1 During the framing of the structure, frame the window opening according to the instructions on page 87. Add extra trimmer studs, if needed, to accommodate the window. Measure the window and draw the perimeter on the inside of the sheathing. Drill small diameter holes through the sheathing and siding at each corner.

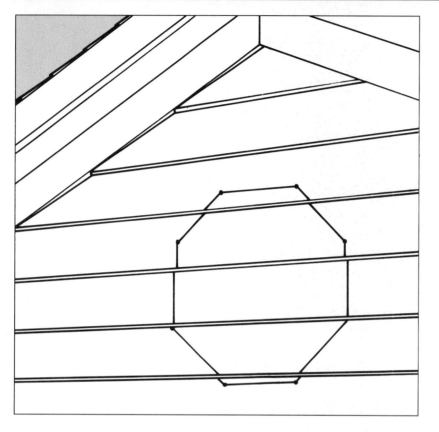

CAUTION

Always wear proper eye protection when using power tools (pages 3 and 4). Also, be sure to keep your hands out of the danger zone (page 5).

2 On the outside of the structure, connect the drilled holes with a chalk or pencil line. Cut out the shape with a sabre or reciprocating saw.

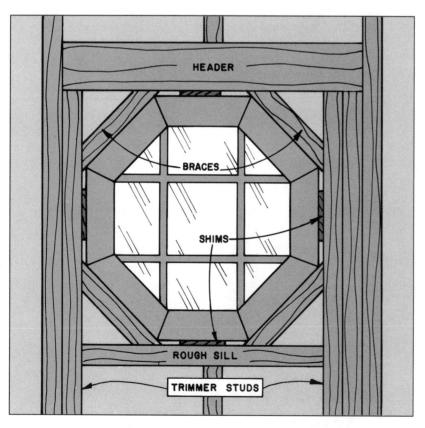

4 Have a helper hold the window in the opening from the outside while you check for level. If needed, drive shims between the frame and the framing members — header, rough sill, trimmer studs and braces. Nail the window in place.

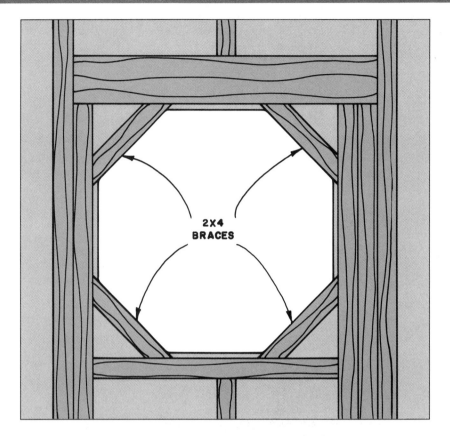

3 Cut and toenail four diagonal 2 x 4 braces in the opening. Cut a piece of flashing the length of the upper half of the window frame and 4 inches wider than its thickness. Cut 4-inch long by 1-inch wide tabs in the flashing. Bend the tabs at right angles, then form the flashing around the frame (page 191). On the outside of the structure, insert the tabs of the flashing between the sheathing and the siding at the top of the opening.

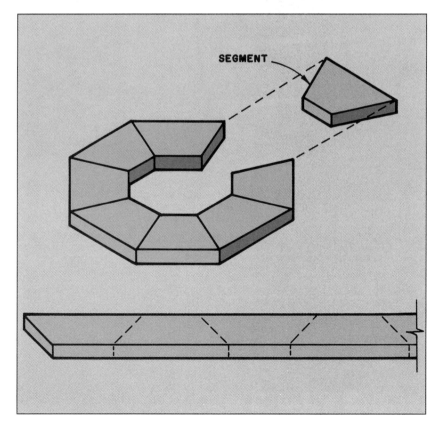

5 Install the exterior trim that came with the window, or make your own decorative trim that matches or complements the other trim on the structure. Caulk where the trim meets the window frame and the siding.

To make octagonal trim, measure and cut the eight segments at an angle of 22½°. Laying out and cutting the segments as shown saves stock and effort.

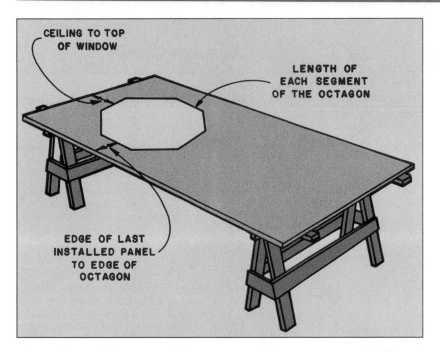

CEILING TO TOP OF WINDOW

LENGTH OF EACH SEGMENT OF THE OCTAGON

EDGE OF LAST INSTALLED PANEL TO EDGE OF OCTAGON

6 During the installation of drywall or paneling you'll need to draw the cutlines for the octagonal window on the panel. To mark the cutlines, you'll need to make three measurements: the distance from the ceiling to the top of the octagon, the distance from the edge of the last installed panel to the edge of the octagon, and the length of each segment of the octagon. Transfer the measurements to the panel. Use a sabre saw to cut out the opening. Install the panel and add the interior trim.

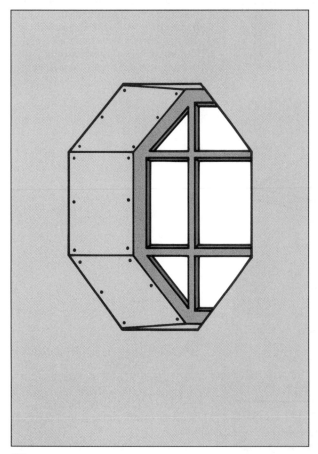

7 The recessed area of the window can be covered with pieces of drywall, paneling or wood. Measure and cut the pieces to fit the recess. Nail the pieces in place. If drywall is used, use tape and joint compound to cover the joints. If paneling or wood is used, add molding to cover the joints.

Semi-circular and quarter-circular windows look good above a square or rectangular window. They are framed and installed similar to round windows except you'll only be working with part of a circle. You'll have to adjust the instructions for a round window accordingly.

Stationary Windows

Not every window needs to be opened. That is the case with stationary windows. They provide light and help to add a special look to your home. These windows can be almost any shape or size. Long thin ones are placed along doorways or on either side of standard windows. In rooms with vaulted ceilings, large rectangular or triangular ones are placed above standard windows and French doors. The possibilities are numerous. When you plan your new home, think about adding several stationary windows for a dramatic effect.

The framing for stationary windows is relatively simple (pages 199-201). You can install window units with frames or just install a sheet of glass — tempered is recommended. Both of these are held in place with strips of stock or molding. An alternative to installing regular glass in stationary windows is to use stained, etched or leaded glass.

Semi-circular and quarter-circular windows are framed and installed similar to round windows (page 189) except you'll only be working with part of a circle. Adjust the instructions accordingly.

Stationary windows add an interesting look to any structure and are relatively easy to install. Three different shapes of stationary windows are shown here.

1 Frame the opening for a stationary window during the framing of the structure. The procedures to follow are similar to the ones for framing a standard window on page 87.

2 Surround the inside of the opening with stock that matches or complements the other trim used on the structure. After the siding is installed, cut and add the exterior trim around the window.

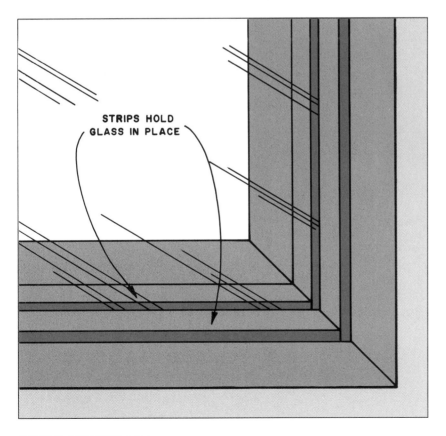

STRIPS HOLD
GLASS IN PLACE

3 The glass in a stationary window is usually centered within the opening. However, it can also be placed toward the front or back of the opening. Use strips of standard stock or decorative molding to form frames that 'sandwich' the piece of glass in position. The width of the strips is not critical and is mostly determined by personal preference and the width of the opening. If the glass is toward the front or back of the opening, you may want to make one of the strips narrower than the other.

Note: You may want to use wooden or metal frames with glass already installed for the stationary window. These are installed between strips of stock or molding just as pieces of glass are.

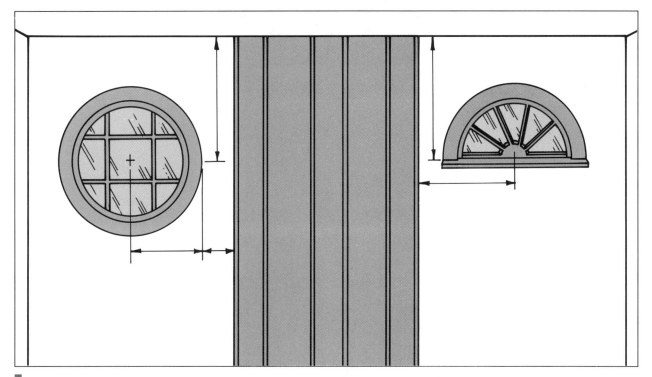

1 To fit drywall or paneling around a circular or semicircular window find the center of the circle. For a circular window, find the point on the window frame nearest to the last installed panel. Measure from the ceiling to this point and subtract ¼ inch for clearance. Then measure from this point to the edge of the last panel installed and to this measurement add the radius of the circle. Use these measurements, to determine the center of the circle on the panel.

For a *semicircular window,* measure from the ceiling to the bottom of the window frame subtracting ¼ inch for clearance; then measure from the center of the bottom of the window frame to the edge of the last panel installed. Use these measurements to determine the center of the base of the semicircle on the panel.

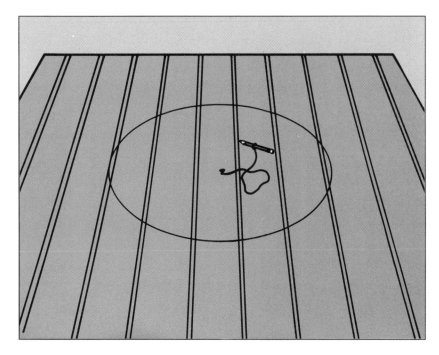

CAUTION

Always wear proper eye protection when using power tools (pages 3 and 4). Also, be sure to keep your hands out of the danger zone (page 5).

2 Drive a nail at the center point on the panel. Tie a string to the nail and attach a pencil on the other end. The length of the string is the radius of the circle or semicircle. Draw the circle, or the semicircle and straight line on the panel. Cut out the opening with a sabre saw.

Note: If the window cannot be covered by one panel, lay two panels side by side as they will be installed. Then draw the window cutline on both panels.

Drywalling or Paneling Around Odd-Shaped Windows

Cutting out the openings for odd-shaped windows in drywall and paneling is not as difficult as you would think. It's simply a matter of transferring the shape and measurements directly to the panel or making a pattern of the window opening. Be sure to take precise measurements.

The shape is cut out of the panel with a sabre saw and then the panel is attached to the wall. Molding will cover any gaps between the window and the panel. Instructions for drywalling around a circular, semi-circular, or other kinds of odd-shaped windows are shown below and on page 202.

Greenhouse Windows

Are you looking for a warm place to start flower seeds in the spring? Or a place to grow an herb garden? Or a sunny spot for your house plants? Or maybe just something to brighten a kitchen? If you are, a greenhouse window is the answer.

You can purchase a prefabricated greenhouse window complete with shelves and a ventilation system. Or, for those of you with a green thumb underneath the sawdust; you can build one.

The construction of the greenhouse window is simple. It is very similar to panel construction for a cabinet carcase on page 146. There are five pieces — top, bottom and three sides — joined together with dowels (page 49) or mortise and tenon joints (page 51). The glass and bottom fit into rabbets (page 45) that are cut in the internal edges. Narrow strips of stock hold the glass in place. A good idea is to hinge the top or one of the small sections for ventilation. Otherwise moisture from the plants could build up on the inside of the glass. Be sure to add a piece of window screen in the section used for ventilation to keep out insects. Choose either single-pane, insulated or thermal pane glass.

It's best to use a wood that will weather well — good choices are cedar, redwood or pressure-treated lumber. You can leave the wood natural or finish it to match other windows. If the window is extremely heavy, make braces and attach them to the underside of the window. Plans for a greenhouse window are on page 204.

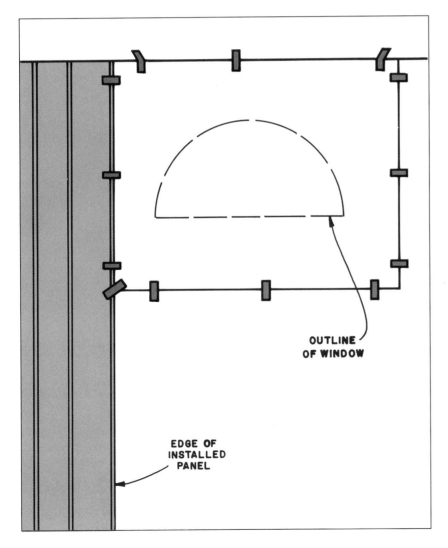

OUTLINE OF WINDOW

EDGE OF INSTALLED PANEL

CAUTION
Always wear proper eye protection when using power tools (pages 3 and 4). Also, be sure to keep your hands out of the danger zone (page 5).

Using a Pattern. Another way to mark the cutlines on a panel for an odd shaped window is to use a pattern. To make the pattern, tape a large sheet of paper over the opening. Align the paper with the top and edge of the installed panel. Press the paper against the opening and draw the outline of the window. Remove the paper and cut out the opening. Position and tape the pattern on the panel, then draw around the opening. Use a sabre saw to cut out the opening.

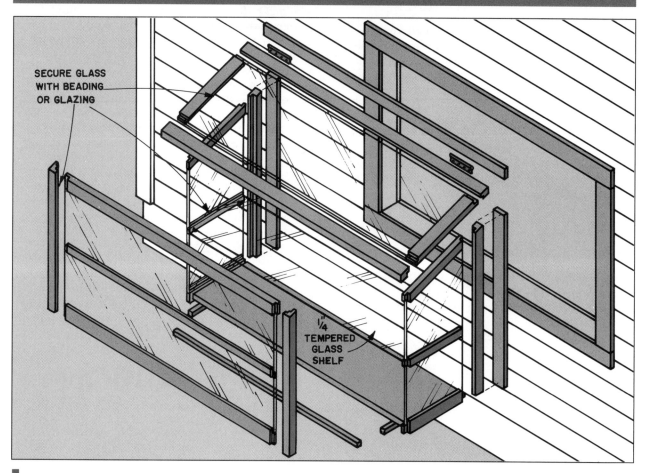

SECURE GLASS
WITH BEADING
OR GLAZING

¼"
TEMPERED
GLASS
SHELF

1 During the framing of the structure, frame the opening for the window according to the instructions for framing a standard window on page 87. Determine the size of your window and purchase lumber — redwood, cedar or pressure-treated lumber are good choices.

Using your measurements and the exploded view above as a guide, cut the stock to size. Cut rabbets in the frame pieces to accommodate the glass and the bottom of the window. Hinge the top for ventilation. Use dowels, or mortise and tenon joints to join the frame pieces. Assemble

the components with waterproof resorcinol glue. Do not install the glass until the window unit is mounted to the structure.

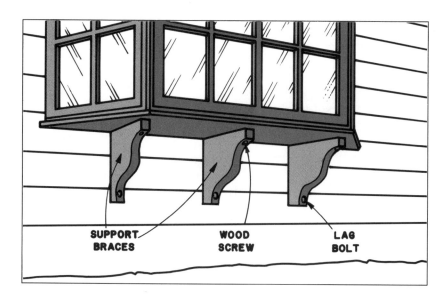

SUPPORT BRACES

WOOD SCREW

LAG BOLT

2 With some helpers, lift the window into place. Mount the window using screws long enough to penetrate at least one inch into the framing members. Generally, no additional support is needed unless the unit is extremely heavy. If it is, add support braces to the underside of the window. Attach the braces with lag bolts into the studs and fasten the tops into the bottom of the window with wood screws. Add exterior trim around the window and caulk between the trim and the window, and between the trim and the siding.

Install the glass and hold it in place by nailing thin strips of stock to the frame. Caulk between the glass and the frame.

Installing a Bay Window

A bay window is quite dramatic. On the outside it gives your house a special look. On the inside it can create a warm and sunny alcove for dining, or by adding a window seat, it becomes a cozy area for reading or storage.

A bay window, like the preceding greenhouse window, is available as a prefabricated aluminum unit, complete with roof. Or you can buy three individual windows and install them as you would a standard window. If you install a prefabricated unit, you'll need several helpers because the unit will be extremely heavy and awkward to lift into place. Be sure to check with your local building department before you start to see if a permit is necessary or if there are any restrictions that may apply.

Whether you use a prefabricated unit or three windows, the framing for a bay window is basically an extension of the framing of the main structure. The main floor joists are extended to create the alcove, walls and windows are framed, and a roof with shingles is added. On pages 206-209 you'll find instructions for framing and installing a bay window.

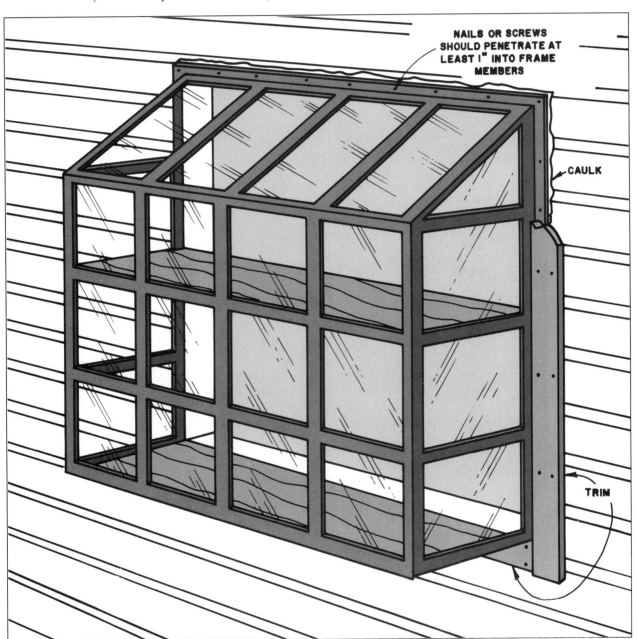

Installing a Window Greenhouse. During the framing of the structure, frame the opening for the greenhouse window unit according to the instructions for framing a standard window on page 87. Install the window unit according to the manufacturer's instructions. Use several helpers to lift the window into place. Check level and plumb, and add shims if necessary. Nail or screw the window to the framing members and caulk all around the window. Generally, metal window units are lightweight so they require no additional supports. After the siding is installed on the structure, add the exterior trim around the window and caulk between the trim and the window and the trim and the siding.

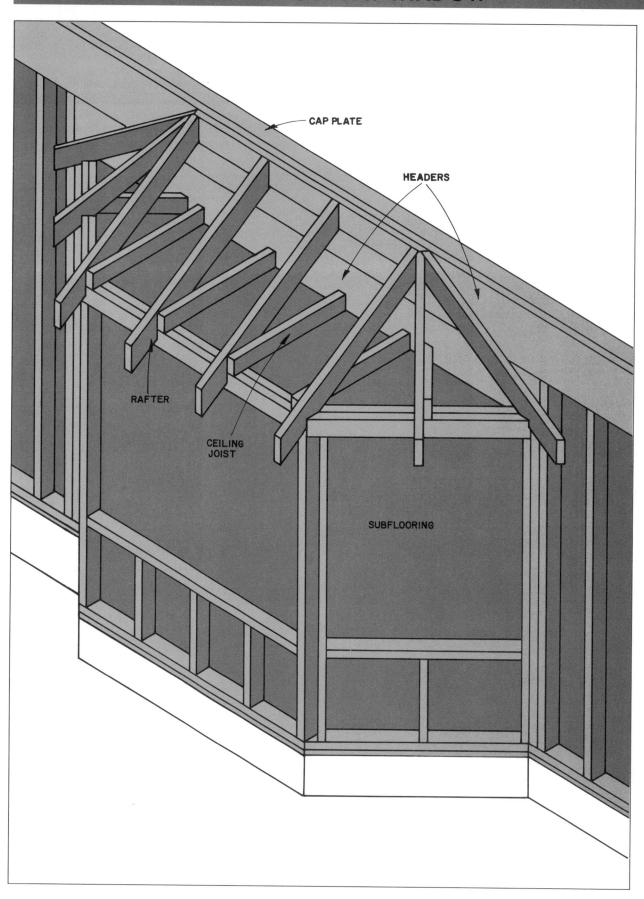

CAP PLATE

HEADERS

RAFTER

CEILING
JOIST

SUBFLOORING

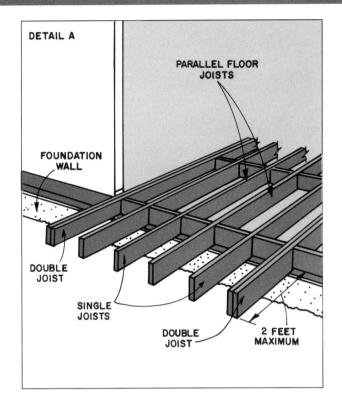

DETAIL A

PARALLEL FLOOR JOISTS

FOUNDATION WALL

DOUBLE JOIST

SINGLE JOISTS

DOUBLE JOIST

2 FEET MAXIMUM

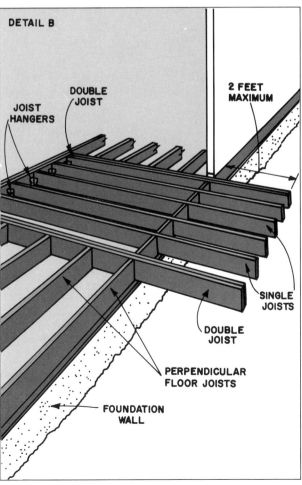

DETAIL B

JOIST HANGERS

DOUBLE JOIST

2 FEET MAXIMUM

SINGLE JOISTS

DOUBLE JOIST

PERPENDICULAR FLOOR JOISTS

FOUNDATION WALL

1 The framing for a bay window is done during the framing of the main structure. Arrange the framing so that the floor joists for the main part of the structure extend (two feet maximum) beyond the foundation wall. The floor joists forming each side of the bay should be doubled so they can carry the necessary load. If the floor joists for the bay window will run parallel to the main floor joists, extend the main floor joists as shown in *Detail A*. If the floor joists will run perpendicular to the main floor joists, frame the floor as shown in *Detail B*.

Instructions for framing the components and installing the subflooring are found in Chapter 6. When the framing is complete, install the three individual windows (pages 114 and 115).

Note: If you are installing a prefabricated bay window unit which usually includes a metal roof, the framing instructions are a little different than the ones above for installing three individual windows. First, frame the area under the window. Next install the window unit following the manufacturer's instructions. Get several helpers to hold the window unit while it is secured to the framing members. Then complete the upper framing and attach the roof.

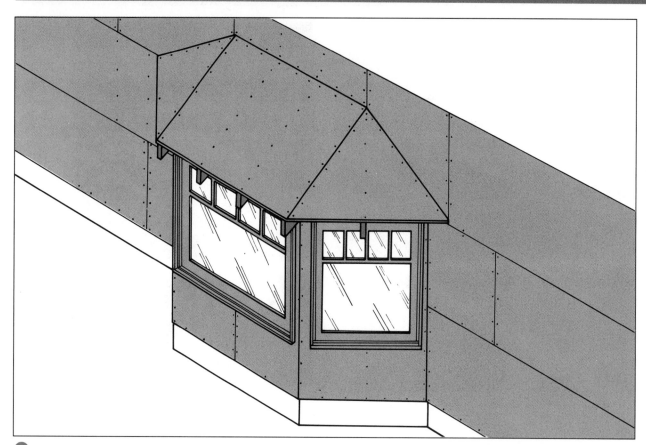

2 Cut and nail sheathing to the underside of the floor joists, to the framing below the windows, and on top of the rafters. Nail siding to the areas below the windows and add trim around the windows.

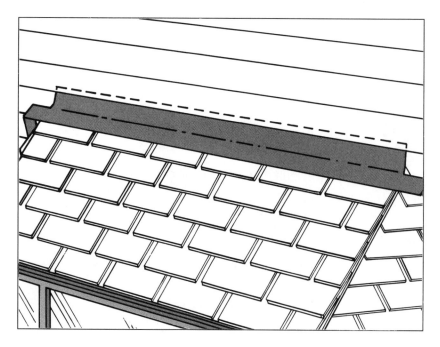

5 Cut a piece of flashing 8 inches longer than the width of the roof. Notch the ends of the flashing and bend the edge at a right angle. Slide the flashing between the siding and the sheathing. Coat the top of the last row of shingles with roofing cement. Press the flashing into the cement. Apply cement to the under-side of the notched ends. Nail the flashing to the roof and bend the notched ends downward. Run a bead of caulk along the gap where the shingles and the siding meet. Add shingles to the hip area of the roof (page 109).

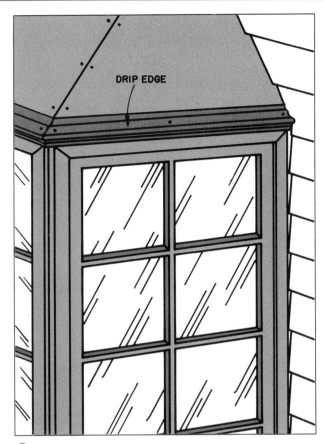

DRIP EDGE

3 After the siding is installed on the main part of the structure, nail a metal drip edge along the bottom edges of the roof sheathing where the sheathing meets the top of the window frame.

4 Install the roofing underlayment and the shingles following the instructions on pages 105 and 108. Cut and bend pieces of flashing at right angles. Starting at the bottom, slide the flashing under each course of shingles and between the siding and the sheathing. Overlap each piece of flashing and nail it at the top.

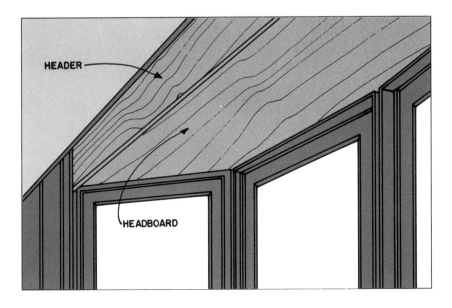

HEADER

HEADBOARD

6 Add insulation between the rafters and ceiling joists. Measure and cut a piece of drywall or veneered plywood for the headboard. Hold the headboard in place and check for level. Add shims if needed and nail the headboard in place. Nail through the shims and into the header. Use molding to cover any gaps between the tops of the windows and the headboard and to add a decorative touch around the perimeter of the opening.

Building a Bay Window Seat

A bay window seat is an attractive addition to any room. It can be a place for house plants, decorator items or a large fancy cushion. Surround it on both sides with built-in bookcases or cabinets for a more decorative look, plus additional storage.

Building a window seat in a bay window is not difficult, as you can see in the illustration on this page.

The simple frame attaches to the framing members of the bay window. Veneered plywood is used for the top of the seat and the front can be veneered plywood, drywall or paneling. Add attractive molding where the top meets the windows and the front, and where the front meets the floor.

Apply the finish of your choice. If the veneered plywood has an attractive grain pattern and is a good grade with no loose knots and cracks, use a natural

BUILDING WINDOW SEATS

A Bay Window Seat. Frame a window seat for a bay window as shown. Generally the height of a window seat is 20". Use veneered plywood for the top of the seat, and veneered plywood or drywall for the front of the seat. Use narrow molding to cover any gaps between the bottoms of the windows and a decorative molding along the front of the seat.

stain and let the grain show. This will give your window seat a rich, warm look.

Creating an Area for a Window Seat

It's not necessary to have a bay window in order to have a handsome window seat. You can build a window seat under almost any window. Just build a frame and cover it with veneered plywood or you can create an area for a window seat by framing the area between a window and the side walls. If you have vaulted ceilings, position a window seat underneath an arched or round window for a dramatic effect. Or if you have a large picture window it can be enhanced with a window seat. A child's room is an excellent place for one. It enables small children to see out without the help of step stools or chairs. An older

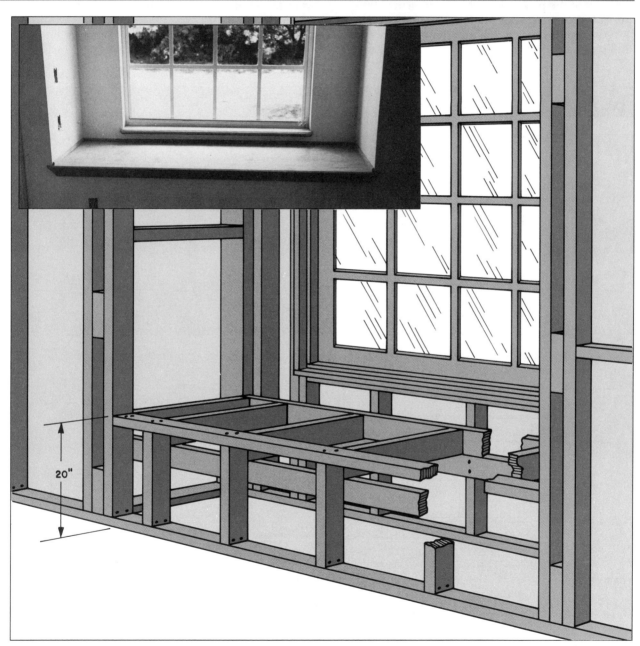

A Standard Window Seat. Generally, you can locate a window seat under most any window in your home. The rectangular seat shown here adds to the elegance of the large cathedral window. The framing for this window seat is merely an extension of the wall framing on both sides of the window. Red oak veneered plywood, matching molding, and drywall were used to cover the simple framing.

child could use a window seat for reading, listening to music or just daydreaming.

To create the area for the window seat, start by adding framing to the wall area on both sides of a standard window. Cover the wall framing with drywall or paneling to match the rest of the room or you can be creative. There are many possibilities — add shelves for books and collectibles; and cabinets, drawers or closets for storage.

The framing and building instructions for this type of seat are the same as those for "Building a Bay Window Seat" on page 210. The only exception is that the framing for the seat is a rectangular box and the ends of the plywood for the seat are cut at 90° instead of at an unusual angle.

Also, on this page you'll find drawings of several different wall arrangements for either or both sides of the window seat. Note the various arrangements of shelves, drawers, cabinets and doors.

Framing and Installing Skylights

A skylight in the vaulted ceiling of a bedroom or bathroom adds light during the day, and a view of the moon at night without taking up wall space or affecting your privacy. Add a skylight in a family room or entrance way and you can grow an indoor garden with trees, plants and flowers. Remember that a skylight should have tinted, insulated or thermal glass to help keep out heat from the sun. Check at your local home improvement store for special shades for skylights.

The skylight frame is attached to the framing members with a curb. If a metal curb is not included

BUILDING WINDOW SEATS/CONT'D.

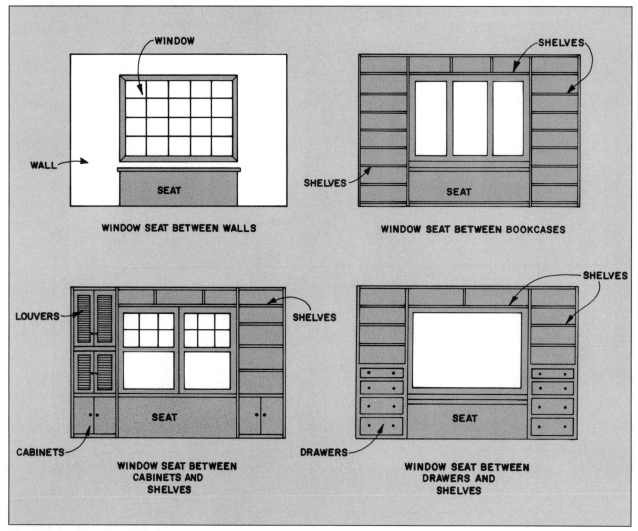

Creating an Area for a Window Seat. You can add a built-in window seat to almost any room by simply framing the area around the window or, better yet, by being creative. Add cabinets, bookcases, shelves or drawers.

with the kit, you will have to make one out of framing lumber. The flashing is placed around the curb and underneath the roofing material to prevent water seepage. Then the curb is nailed to the framing and any gaps and exposed hardware are caulked.

You can purchase preframed tinted skylight kits at home improvement centers and building supply stores. You should be able to install the skylight in a new structure yourself. Just do it before the shingles have been installed on the roof and before the insulation and drywall are added to the ceiling.

If your skylight doesn't come with base and head flashing, have it made at a sheet metal shop. Do not try to make the flashing yourself unless you are familiar with metal working and have the proper equipment. It is very important that the flashing be made properly to avoid leakage.

There are two types of skylight installations. One type is quite complicated, in that it involves the building of a light shaft that funnels the light from the roof, through the attic and into the ceiling of the room. The light shaft requires framing that is covered with drywall. The other type is a skylight that is easily installed in a vaulted ceiling. Instructions for framing a skylight in a vaulted ceiling begin on this page.

If you need to build a light shaft for your skylight, frame the roof opening the same as for a vaulted ceiling. Then add a box type framing between the roof and ceiling opening. Refer to pages 216-217 for instructions that tell how to frame the opening in the ceiling. Also follow the manufacturer's instructions that came with the window.

Check with your local building department; usually this type of work calls for a building permit.

INSTALLING A SKYLIGHT IN A VAULTED CEILING

Before you get started, read the following instructions and the manufacturer's instructions that came with the skylight. Also check your local building department for any restrictions that may apply.

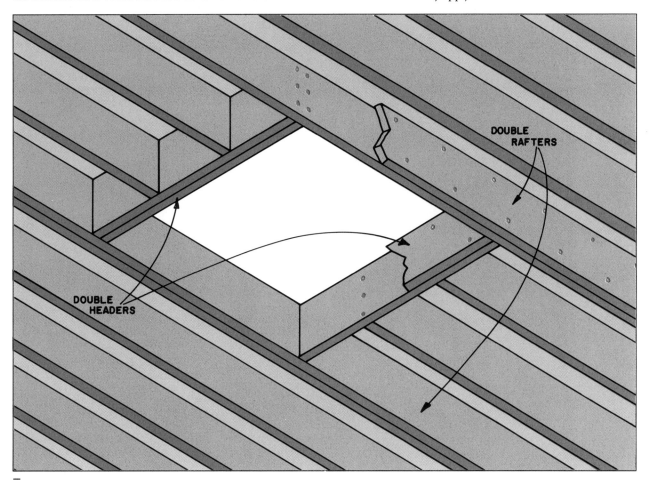

1 Frame the opening for the skylight during the framing of the roof (page 91). Use double rafters and headers to reinforce the framing members as shown.

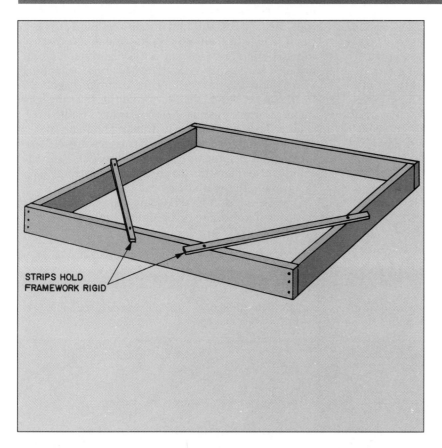

STRIPS HOLD
FRAMEWORK RIGID

2 Next, you need to build a curb. Measure the interior of the skylight frame. This measurement minus ⅜ inch is the outside dimension of the curb. Measure carefully, cut and assemble the curb. Nail strips of scrap stock across the corners to prevent the curb from flexing when it is moved. Remove the scrap stock after the curb is set in place on the roof.

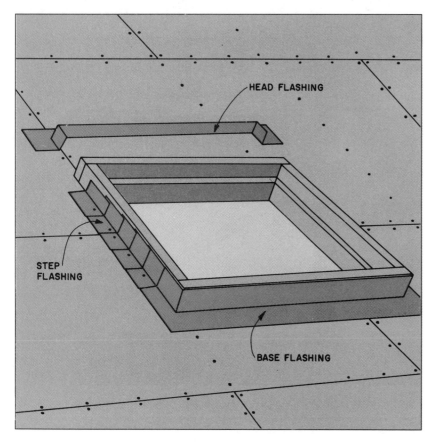

HEAD FLASHING

STEP
FLASHING

BASE FLASHING

4 Install the flashing very carefully in the following sequence: base flashing, step flashing, and head flashing. Use roofing cement or caulk to seal the flashing where it meets the sheathing, the other flashing and the curb. Also cover any exposed nail heads.

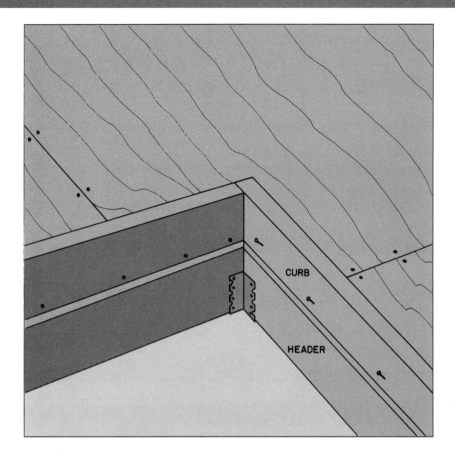

CURB

HEADER

3 Place the curb around the opening. Use a carpenter's square to check that it is square, then toenail the curb to the framing members.

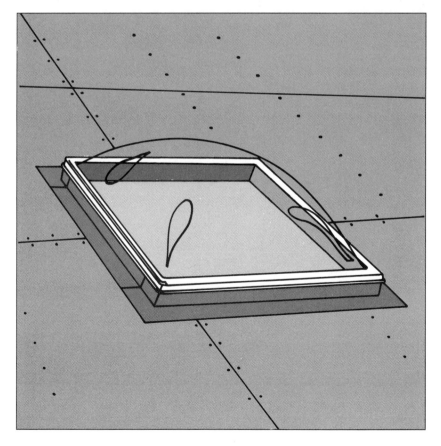

5 Apply a generous amount of caulk to the top edges of the curb. Position the skylight on the curb and fasten it securely. Add caulking where the skylight and the curb meet.

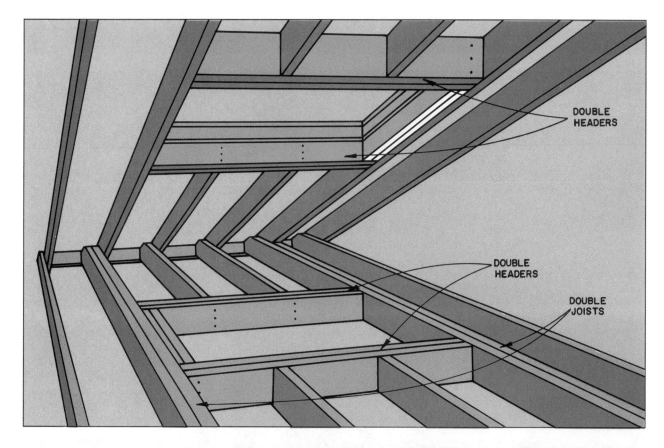

DOUBLE
HEADERS

DOUBLE
HEADERS

DOUBLE
JOISTS

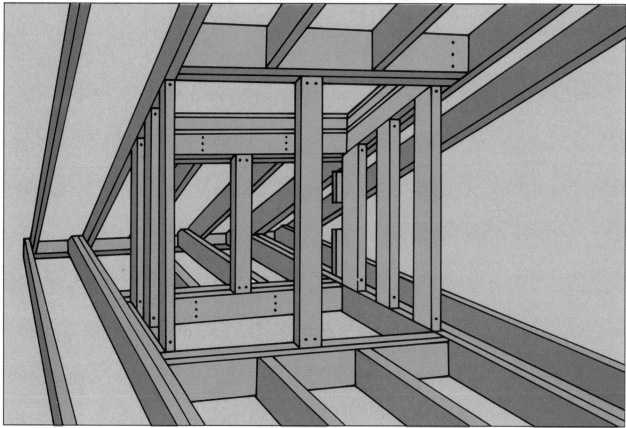

Building a Light Shaft. The framing for a light shaft between a roof and a ceiling is shown here. Use 2 x 4 lumber and follow the instructions on pages 79 and 213 to frame the ceiling and roof openings. Use drywall or paneling to cover the framing.

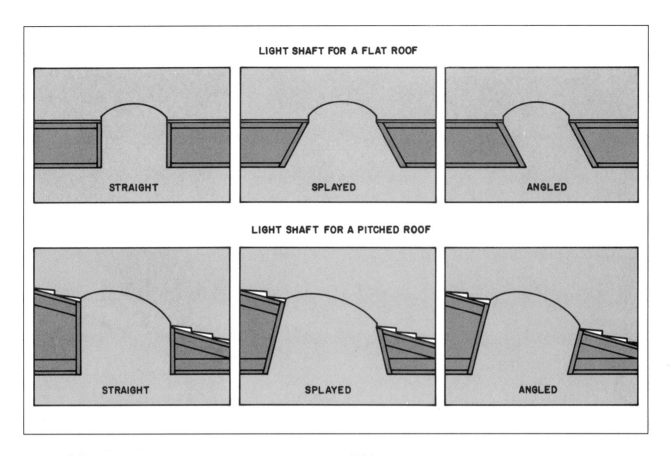

LIGHT SHAFT FOR A FLAT ROOF

STRAIGHT

SPLAYED

ANGLED

LIGHT SHAFT FOR A PITCHED ROOF

STRAIGHT

SPLAYED

ANGLED

Skylights can be installed in flat or pitched roofs, and be straight, splayed or angled. Adjust the framing instructions to accommodate different configurations of light shafts.

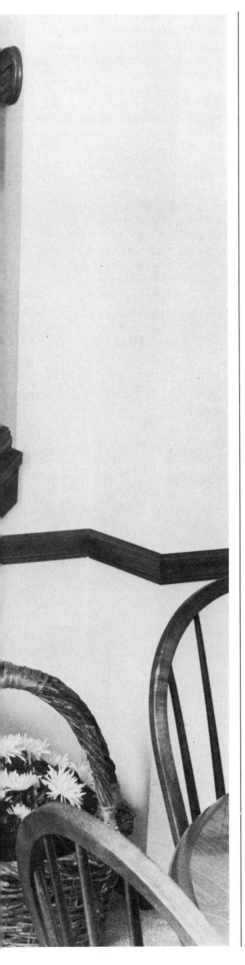

12

Staircases and Railings

Staircases lead from one floor level to another. They are mainly functional; however, the main staircase in your home should also be attractive. If you're building an open staircase, add delicately turned spindles and ornate posts. Also round the edges of the handrail to soften its lines, and add custom moldings on the edges of each stair and where the treads and risers meet. All of these special touches help to give your home a luxurious and unique appearance.

Basic stairs and railings for decks, porches and basements are generally more functional than decorative. Therefore simple construction with few components, very basic joinery, and inexpensive materials can be used. Generally, you can use woods such as cedar, redwood and pressure-treated lumber for outdoor stairs, and an inexpensive grade of softwood for inside ones.

Stairways without walls or with just one wall are considered to be open. A half-open stairway is a combination of both, and a closed stairway has walls on both sides. A stairway can be one flight, or more than one flight with a landing in between.

In this chapter you'll learn how staircases and railings are built and how to install them.

Designing a Staircase

A typical staircase consists primarily of stringers, risers, treads and a railing assembly. The stringers provide the main support for the staircase. The treads and risers are mounted to the stringers to form the walking surface and to enclose the back of each stair. Usually a railing assembly is required for safety, so be sure to check your local building code before you get started. The parts of a typical open staircase are shown on this page.

To help you design your staircase, look through books and magazines for plans and ideas that you can adapt for your use. Dress up your main staircase with warm, rich looking hardwoods and custom touches like turned spindles and a handsome railing.

Staircases must be soundly built with the safety and comfort of the user always in mind. When correctly designed, the stairs will have the proper tread width and riser height. Stairs that are too steep or too shallow are dangerous and uncomfortable. Those with too wide a tread and too short a riser will cause the user to lean backward while going up and forward while coming down.

The width of the treads and the height of the risers are strictly regulated. Check your local building department. When the height of the riser and the width of the tread are added together, the total should be between 17 and 18 inches. Thus if the riser is 7 inches high, the tread will be 10 to 11 inches wide. The risers should be between 7 and 8 inches high.

Main staircases must be wide enough for at least two people to pass comfortably and to permit furniture to be carried up or down. The accepted minimum width for a main stairway is 3 feet, but 3 feet 6 inches is preferred. The handrail height found to be the most comfortable is 30 inches up from the treads and 34 inches from the floor on the landing.

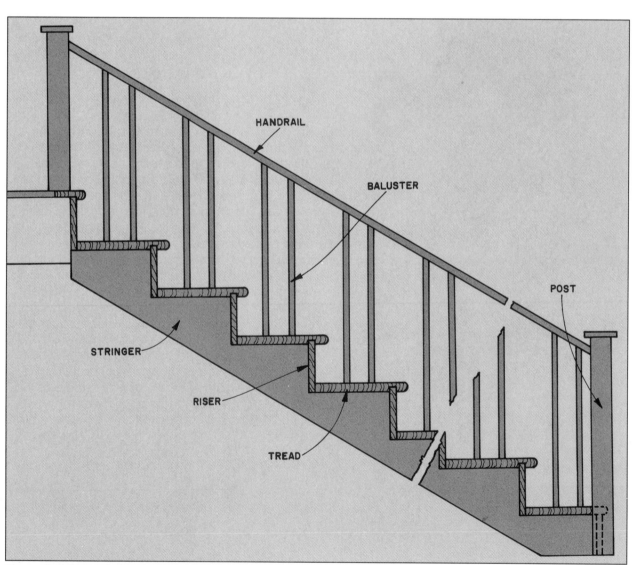

Parts of an Open Staircase.   This typical open staircase includes stringers, risers, treads and a railing assembly (handrail, balusters and posts). Build on these basics to create your own staircase.

Building a Staircase. The procedures for constructing staircases are as simple or complex as their design and intended use.

You can build a utility staircase for a basement, porch or deck with basic woodworking skills, a few tools and some inexpensive lumber. Components can consist of notched stringers, treads, and a railing. Risers are optional.

An even simpler method for building utility stairs is shown on this page. As you can see, it's not always necessary to notch the stringers to support the treads. Just nail cleats to the stringers and then nail treads to the cleats. Another simple method is to cut dadoes across the surface of the stringer and glue and nail the treads in place. Add a simple railing to comply to your local code.

Building an attractive open staircase requires considerable work and skill. You'll need to know joinery techniques, and how to turn spindles and make molding. The stock you use — hardwood or softwood — depends on the finish to be applied, how visible the staircase will be, and how much you want to spend.

Alternatives to building all the components yourself are to purchase spindles, molding and pre-cut treads and risers; or to purchase a staircase kit from a building supply store and install it yourself.

You can dress up the staircase in various ways with custom-made molding and special touches as desired. The treads can overhang or be flush with the edges of the risers. When the treads will be completely visible or partially covered with a strip of carpet you can round over the front edge of each tread for a smoother, cleaner appearance.

Use the instructions for building a typical open staircase that appear on pages 222-223 as a guide to build your staircase.

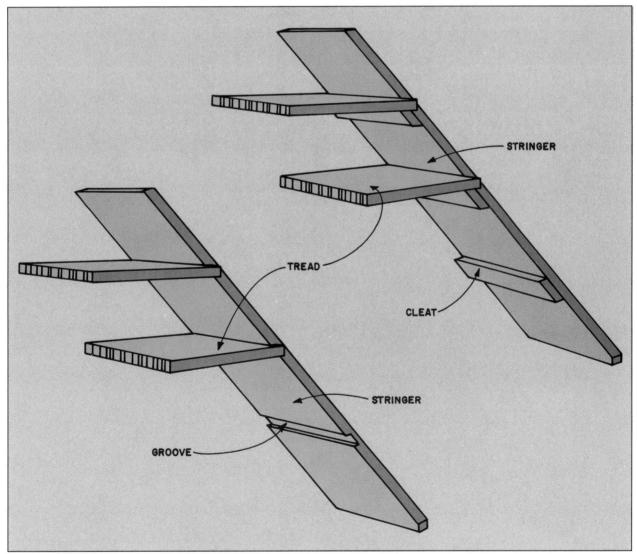

Simple Stairs. Here are two types of stairs that are not very difficult to build. Instead of using the standard notched stringer, use a piece of 2 x 12 stock. Add cleats or cut grooves in the surface to support the treads. In most cases, risers are not necessary. These types of stairs are used for decks, porches and basements. Add a simple railing for safety.

BUILDING AN OPEN STAIRCASE

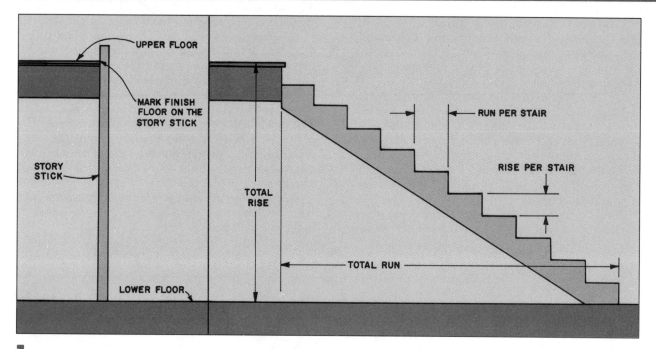

1 To calculate the rise and run of a stairway, first find the total rise of the staircase. Stand a straight piece of lumber (story stick) between the upper and lower finish floors. Mark the finish floor on the story stick. This measurement is the total rise.

Next divide the total rise by 7 to obtain the number of stairs. Then divide the number of stairs into the total rise to determine the rise per stair.

Here is an example of calculating a run for a total rise of 110.5 inches: $110.5 \div 7 = 15.78$ risers. Therefore, you'll need either 15 or 16 risers. If you choose 15, each riser will be more than 7 inches. If you choose 16, each riser will be less than 7 inches. Divide both 15 and 16 into 110.5"

and choose the number which is closest to 7. $110.5" \div 15 = 7.37"$ ($7\frac{3}{8}"$); $110.5" \div 16 = 6.91$ ($6\frac{29}{32}"$). The run for each stair is $6\frac{29}{32}$.

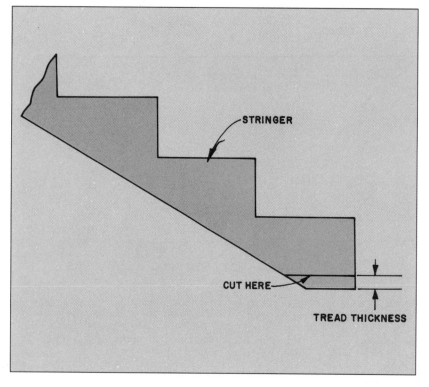

4 Cut out the stringers. Since the tread thickness will add to the height of the first step and diminish the height of the last step, measure the thickness of the tread and cut this amount off the bottom of the stringer.

STAIRCASES AND RAILINGS

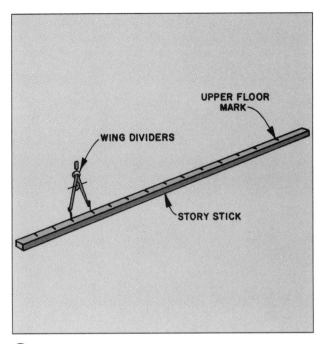

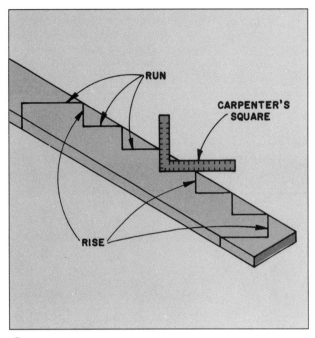

2 Set 6²⁹⁄₃₂ on a pair of wing dividers and step off this distance on the story stick from the upper floor mark to the bottom floor line. If the last step of the wing dividers doesn't end exactly on the bottom floor line, adjust them slightly and repeat until it does. When it comes out exactly, the dividers are set at the height of each riser.

3 To lay out the stringers, mark the height of each riser across the tongue of a carpenter's square with a grease pencil. Mark the width of the tread across the blade. Lay the square on the stringer stock and mark the rise and run as shown.

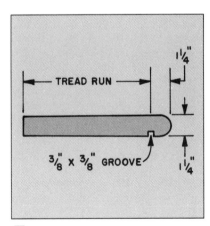

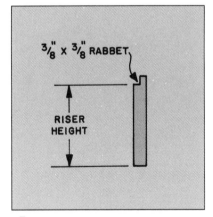

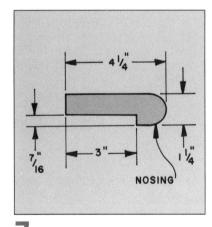

5 To make the treads, plane the stock to 1¼ inches and cut the treads to length. Round the front edges of each tread using a shaper or router. Then rip the treads to the exact width — tread run plus overhang. Cut a ⅜ inch x ⅜ inch groove, the full length of each tread, 1¼ inches back from the front edge. This groove will accommodate the top edge of the riser.

6 Plane the stock for the risers to ¾ inch and cut it to length. Cut a ⅜ inch x ⅜ inch rabbet along one edge of each riser. Rip each riser to width — ⅜ inch wider than the height of the riser + tread thickness. The top riser should be ¾ inch wider than the rest.

7 Plane the stock for the nosing to 1¼ inches and cut it to length. Shape one edge to a half circle. Rip the stock to 4¼ inches wide and cut a 3-inches-wide by 7⁄16-inch-deep rabbet in the surface.

INSTALLING AN OPEN STAIRCASE

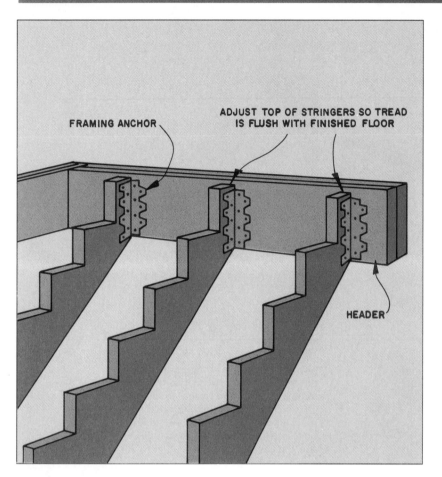

FRAMING ANCHOR

ADJUST TOP OF STRINGERS SO TREAD IS FLUSH WITH FINISHED FLOOR

HEADER

1 Position the top of the stringers so the tread, when installed, will be flush with the finish floor. Then attach the stringers to the header with framing anchors and nail the stringers to the wall studs.

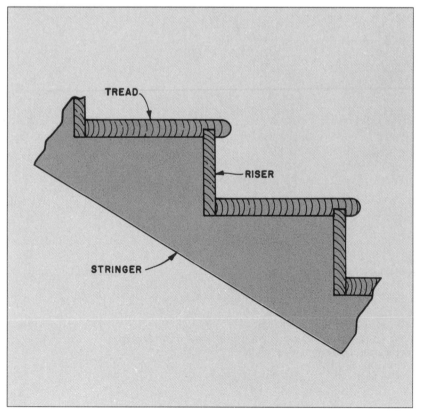

TREAD

RISER

STRINGER

3 Nail the risers, rabbeted edge up, to the stringers using finishing nails. Place the rabbeted edge of the risers into the groove in the bottom of the treads. Nail the treads in place using finishing nails. From under the stringers, nail through the back surface of the risers into the back edges of the treads.

STAIRCASES AND RAILINGS

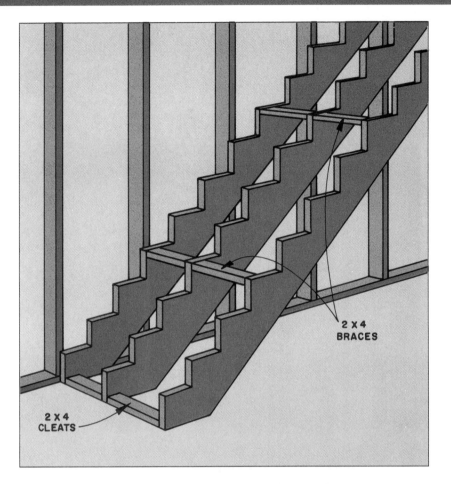

2 X 4
BRACES

2 X 4
CLEATS

2 One-third of the way up the stairs, toenail and/or face-nail two spacers between the stringers. Repeat the procedure two-thirds of the way up the stairs. Then toenail two cleats to the bottom of the stringers and the floor.

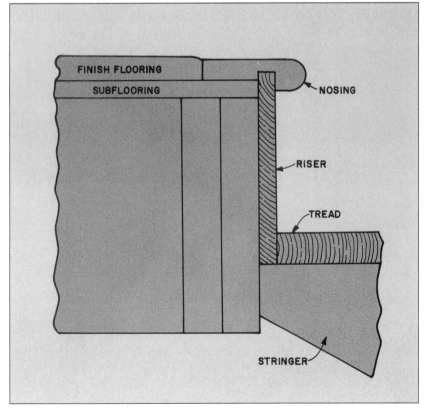

FINISH FLOORING

SUBFLOORING

NOSING

RISER

TREAD

STRINGER

4 After the finish floor is installed, nail the nosing to the subflooring using finishing nails. Set all the nail heads with a nail set.

Installing a Typical Open Staircase. Frame the opening for the ceiling where the stringers will be mounted according to the instructions for framing a floor opening on page 79. Then install the open staircase following the detailed procedures on pages 224-225.

After the staircase has been installed, frame the areas around the top and underneath your staircase as shown on this page. You can then cover the framing with drywall or paneling.

Building the Railing Assembly

Normally, stairs have railings for safety. This is governed by their location, their function and the local building codes. Some railings are more elaborate than others. The railing assembly for your main staircase should be made with hardwood that matches or complements the other trim in your home. Railing assemblies for outdoor use should be made out of lumber that can withstand the weather.

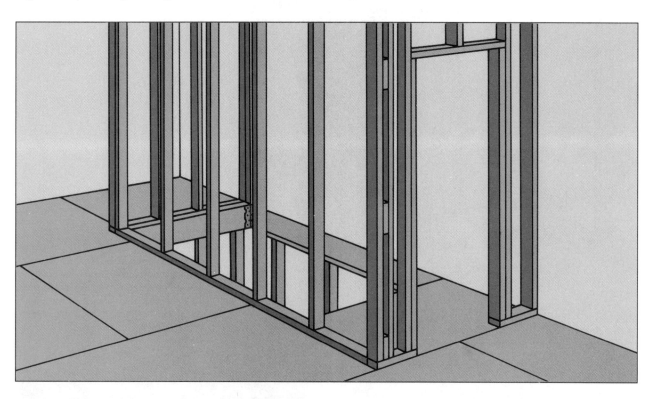

Framing the Stairwell. The ceiling opening was framed during the framing of the structure (page 79). After the staircase is installed, frame the areas around the top and underneath your staircase. The framing process is very similar to standard wall framing explained on pages 84 and 85. When the framing is complete, add the drywall according to the instructions on pages 126-127.

The handrail, balusters and posts are the common parts of the railing assembly. The assembly is attached directly to the stringers or a middle rail. The handrail should be continuous from one floor level to the next whenever possible. Since it will be difficult to find one long perfect piece of lumber, you'll probably have to join one or more pieces together. Also the handrail should be smooth, without sharp edges and free of splinters. So it's best to round over or shape the edges using a router or shaper. The balusters can be square, flat, or round. The posts must be sturdy and securely attached. Guidelines for building a common railing assembly appear on this page.

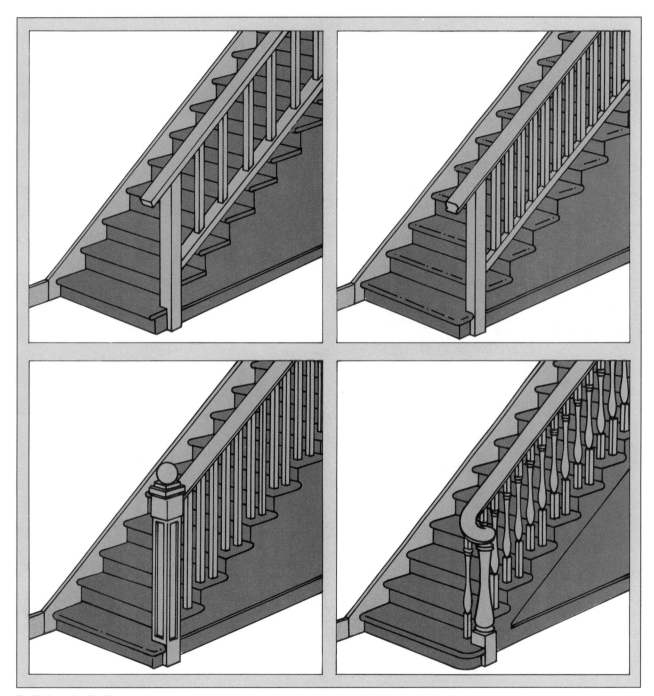

Building the Railing.   The handrail, balusters and posts are the common parts of the railing assembly. Several simple and elaborate railings are shown here.

Make the handrail and middle rail (if used) out of one piece of stock. Two ways to secure the balusters are: Cut a groove in the underside of the handrail to accommodate the ends of the square balusters, or drill holes to accept the spindles.

Cut the square balusters at an angle on both ends so they fit against the underside of the handrail and the surface of the middle rail. If you are making spindles, turn a dowel on both ends to fit into drilled holes in the handrail and middle rail.

Use 4 x 4 lumber or glued-up stock for the posts. Leave the stock square, or turn it on the lathe for a more decorative post.

13

Open Decks

If you're an outdoor enthusiast, or if you simply enjoy a cookout with a few friends, adding an open deck to your home may be a wise move. Not only will a deck increase the value of your property, it'll extend your living space.

You can add a deck outside a family room at ground level, onto an existing patio, or to a bedroom at an upper level of your home. Or, just build a stand-alone deck in a favorite spot in your yard. Depending on the height of the deck, several stairs or a staircase may be necessary to reach the deck from outside.

Open decks are relatively easy to build, require a minimum investment in materials, and most can be constructed in a weekend or two. Your deck can be as simple or elaborate as you desire. You're only limited by the available space and your own skills.

In this chapter you will learn basic deck construction. There are plans that you can use as a guide to design and build your own deck.

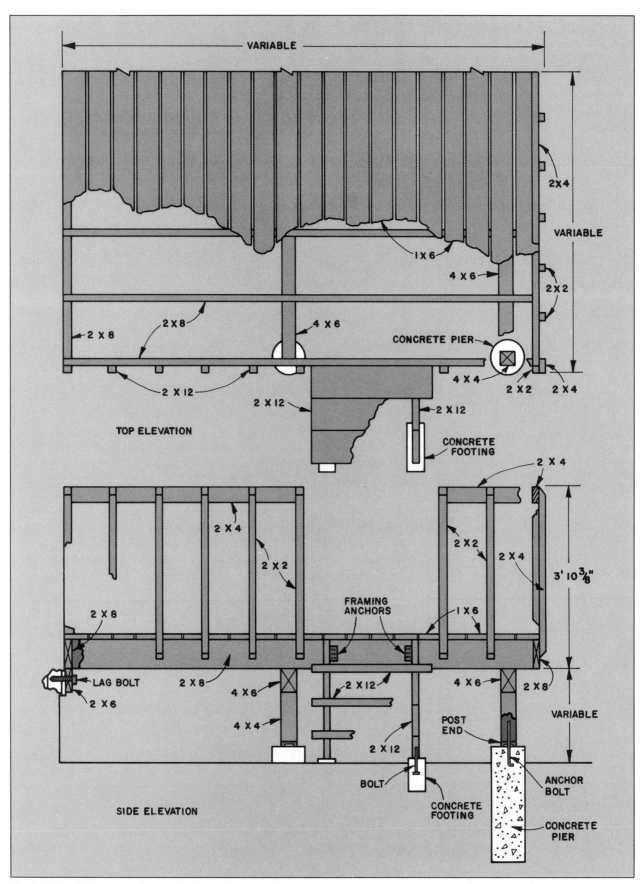

Typical Deck Construction. Use these elevation drawings as a guide to draw the plans and order the materials you'll need to build your deck.

OPEN DECKS

Decks — Additions to Your Living Space

You can increase the living area of your home dramatically by adding a deck. Most decks are located outside of family rooms, living rooms and kitchens because they are generally considered extensions of eating and entertaining areas. However, a deck outside a bedroom would be a great place to locate a hot tub or to just relax. Normally, access from inside your home to both of these decks would be gained through sliding glass or French doors.

If an add-on deck is not possible because your family room, bedroom, or other rooms of your home do not have doors to the outside, you can still have a deck. You can either cut an opening in a wall for a doorway, or build a stand-alone deck in your yard.

Designing the Deck

Design your deck to suit your needs. It can be any size, shape, or height. The area available in your yard, your woodworking skills and the time you have to spend building the deck will usually determine its size. Also be sure to check your local building code for any restrictions that may apply.

As far as shape goes, your deck can be merely a small square, simple rectangle, or an elaborate octagon. The height is determined by the terrain of your yard and at what level of your home the deck will be added.

The height of your deck determines the number of stairs and if a railing is needed. If yours is slightly above ground level, one or two steps should suffice.

If your deck is located at an upper level, you'll need a staircase.

The typical deck construction shown on page 230 is quite versatile; it can be adapted easily to suit your tastes and needs. When you draw the plans for your deck, try to use dimensions that are divisible by 2′ — otherwise you'll be cutting standard stock to size and that wastes time and money. And once again, remember to check your local building codes before you get started.

Building the Deck

Decks, stairs and railings are usually built with simple joinery and basic woodworking techniques. Even the materials are simple. Lumber with knots and minor imperfections is acceptable. The best choice of building materials is douglas fir, redwood, cedar, or pressure-treated lumber. These woods weather well. They require no finish, but they can be stained to match or complement the other trim on your home if you wish.

The pier-and-post foundation is perfect for all types of terrain — simply adjust the length of the posts to compensate for the slope. Build the deck close to the ground or raise it up to make a second-story balcony.

You will find step-by-step instructions for building a simple deck beginning on page 231. Included in these instructions are many valuable tips that will help your deck withstand the elements. It's important to construct your deck so that wooden components will not sit in standing water, or the ends of the stock are not directly exposed to water. Otherwise the wood will rot, thus shortening the life of your deck.

BUILDING A DECK

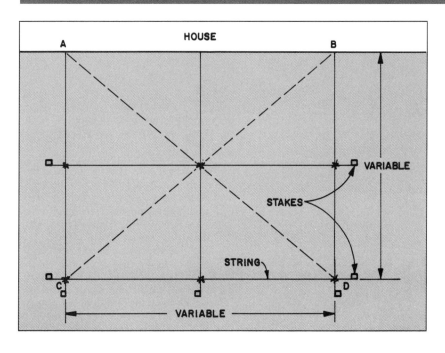

1 Use stakes and string to lay out your deck. Stretch the string between the house and the stakes. Use a string level to make sure the strings are perfectly horizontal. Check that your layout is square by measuring diagonally from corner to corner. AD must equal BC. Then find the locations of the piers by measuring along the string. Locate the center of each pier no more than 6 feet apart. Dig the holes for the piers 24 to 36 inches deep, or below the frostline for your area. Set 8-inch diameter cardboard forms in the holes. (You can also use 8-inch stovepipe.) The tops of the forms should be at least 4 inches above ground level. Place 2 to 3 inches of gravel into the bottom of each form to help drain ground water away from the piers.

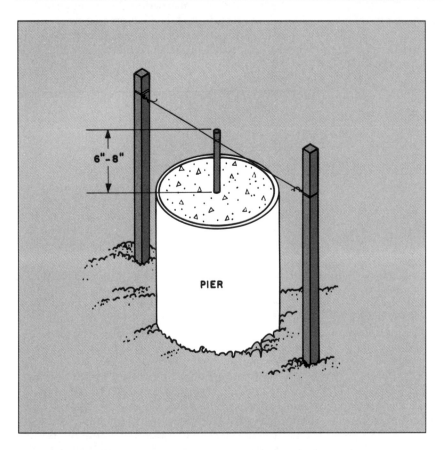

2 Mix the cement and pour it into the forms. Before the cement cures, position the anchor bolts in the center of each pier. To hold the bolts at the proper height, drive two small stakes on each side of the piers. Wrap a wire around the ends of each bolt, then wrap the ends of the wire around the stakes so that the bolts are suspended in the wet cement. The tops of the bolts should protrude 6 to 8 inches. Wait at least 24 hours for the cement to cure, then remove the forms.

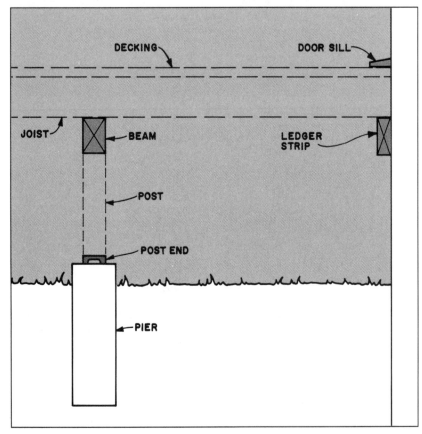

4 With string and a string level, carefully calculate the length of each post. The posts must support the beams so that the top edge of the beams is level with the top edge of the ledger strip. Drive stakes next to each pier. Attach a string to the top edge of the ledger strip, stretch it out over a pier, and attach the end to the stake. Use a string level to check that the string is perfectly horizontal. Measure the distance from the string to the pier. From this distance subtract the width of the beam, and the thickness of the metal post end. This will give you the length of the post for that pier. Repeat this process to find the length of each post.

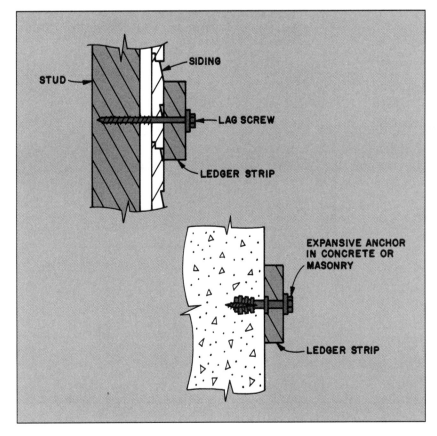

3 Calculate where you want to attach the ledger strip to your house. You'll want the surface of the deck positioned just below the door sill that will lead from the house to the deck. Add the thickness of the decking and the width of the joists. This total is the distance from the bottom of the door sill to the top edge of the ledger strip. Snap a chalk line along the wall of the house where you want to attach the ledger strip.

Cut a 2 x 6 ledger strip and place it even with the chalk mark. Check for level, then attach the ledger strip permanently to the house. To secure the ledger strip to a wood frame member, use lag screws (shown on the left). Make sure each screw sinks into a stud, otherwise the ledger will not support the weight of the deck. To secure the ledger strip in concrete or masonry, use expansive lead anchors (shown on the right).

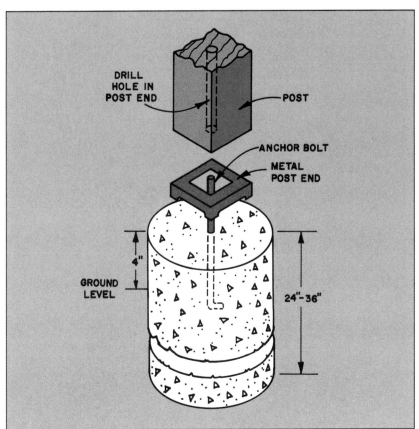

5 Cut the posts to their proper length, and drill holes for the anchor bolts in the lower end. Make the holes $\frac{1}{16}$ inch larger than the diameter of the bolts. This will give the wood some room to expand during wet weather. To keep standing water from rotting out the posts, use metal post ends to raise the posts off the surface of the pier. Install all the posts.

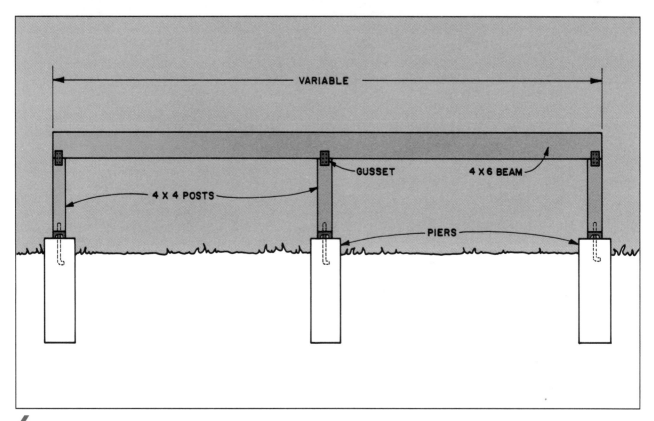

6 Cut the 4 x 6 beams. If you can't get 4 x 6's long enough, overlap two shorter beams. Plan this overlap so that it occurs over a post. Tack the beams in place, then use a carpenter's level to check for level. Attach the beams to the posts with gussets.

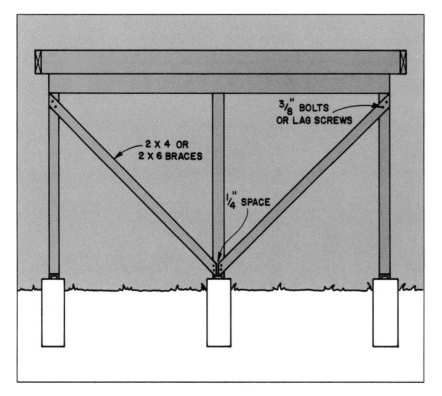

8 Check your local building codes to determine whether you must install braces. (A rule of thumb is that if a post is over 2 feet long, it's a good idea to brace it.) Cut 2 x 4's or 2 x 6's to fit diagonally from the bottom of one post to the top of another. Miter the ends of the braces so that the end grain will not be exposed to water. Attach the braces to the posts, leaving a ¼-inch space between the braces so that water doesn't collect there. Use ⅜-inch bolts or lag screws to attach the braces to the posts.

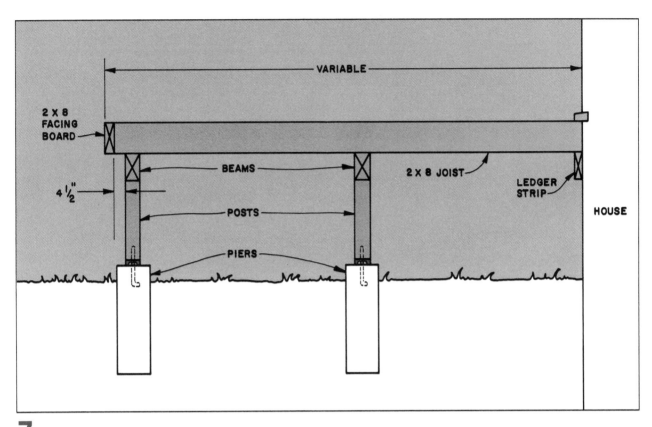

7 Measure for the joists, allowing for a 4½-inch overhang. Cut the joists from 2 x 8 stock. Place the joists on edge over the ledger strip and beams. Be sure the joists are spaced 16 inches on center. Toenail the joists to the ledger strip and beams using 16d nails, or secure the joists in place using joist hangers. Cut a 2 x 8 facing board the same length as the ledger strip. Nail the facing board to the ends of the joists using 12d nails.

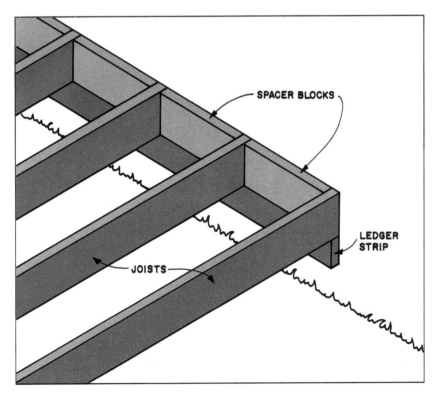

9 Cut spacer blocks from 2 x 8 lumber to provide a nailing surface for the first row of decking and to help keep the joists in position. Fit the spacer blocks between the joists and place them against the house. The spacer blocks must be flush with the top of the joists. Attach the spacer blocks to the joists and the house with 16d nails. If you're attaching the spacer blocks to a masonry wall, use masonry nails.

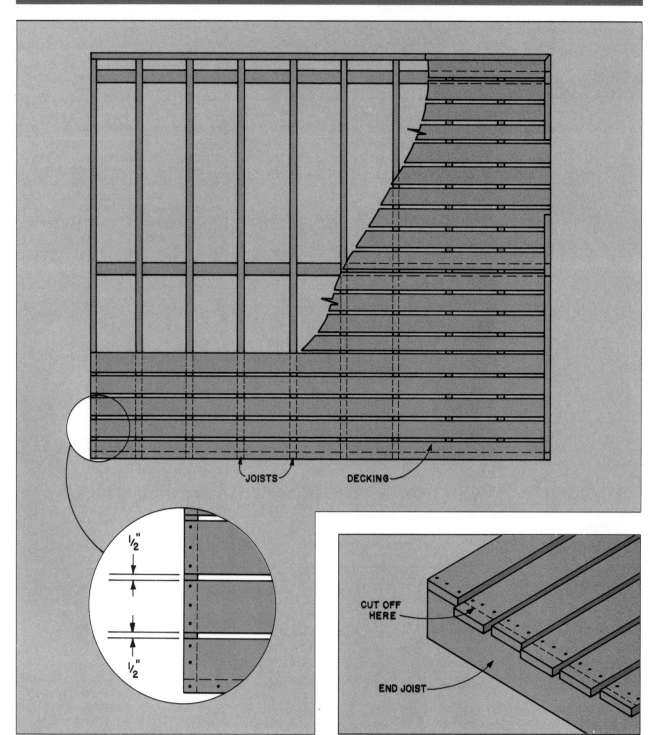

10 Lay the decking in place perpendicular to the joists. Leave a ½-inch space between each piece of decking to allow for expansion in wet weather. Use 12d square-shanked spiral nails to attach the decking to the joists — these nails keep the boards from working loose.

11 Cut the decking to length after you've nailed it in place. While you're installing it, let each piece overhang 1 to 3 inches. Then snap a chalk line even with the outside edge of the end joists. Cut along the mark with a circular saw. To cut the decking close to the house, use a sabre saw or hand saw.

OPEN DECKS

Building Stairs and Railings

The stairs and railings for decks are usually simple in design and are needed primarily for safety. Be sure to check your local building code for standard specifications and any restrictions that might apply. If you need to add stairs and railings to your deck, follow the instructions on pages 220 and 221 to design, build and install them. The stairs and railings are made out of the same stock recommended for the other components of the deck — redwood, cedar, douglas fir, and pressure-treated lumber.

Extra Touches for Your Deck

You can make your deck more attractive by using turned spindles instead of square or flat balusters; by rounding over the edges of the stair treads and the railings; and by adding built-in benches and planters. You could even add a hot tub with a privacy screen. Another extra that you might consider would be to add a cover over all or part of the deck. A cover would provide privacy, plus shelter from the rain and sun. Refer to magazines and books from your local library or bookstore for plans and ideas for these extras.

BUILDING AND INSTALLING THE STAIRS

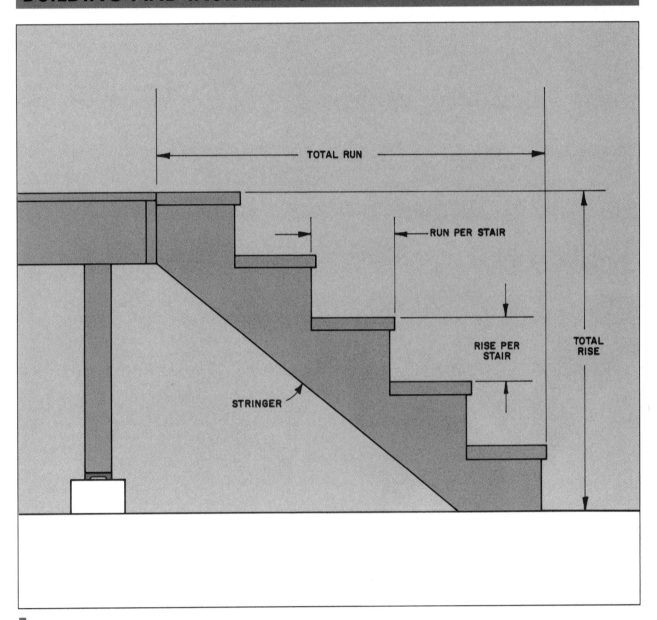

TOTAL RUN

RUN PER STAIR

RISE PER STAIR

TOTAL RISE

STRINGER

1 Calculate the rise and run of the stairs, and cut the stringers from 2 x 12 stock according to the procedures on pages 222 and 223.

You'll need to place the stringers no more than 24 inches apart. If your stairs are wider than 24 inches, make at least three stringers.

OPEN DECKS

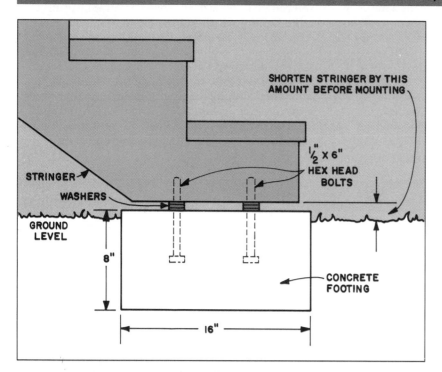

SHORTEN STRINGER BY THIS AMOUNT BEFORE MOUNTING

STRINGER

WASHERS

GROUND LEVEL

½" x 6" HEX HEAD BOLTS

8"

16"

CONCRETE FOOTING

2 The stairs must rest on solid ground or they won't remain level and safe. So, pour concrete footings or use cement blocks to rest the stringers on. Place hex head bolts in the concrete before it sets up. (If you're using cement blocks, fill the holes with cement to set the bolts.) Cut enough stock off the bottom of the stringers to compensate for the washers and the height of the footings above ground level. Otherwise the bottom stair will be too high.

Put 3-4 washers over the ends of the bolts to keep the stringers from sitting in any water that might accumulate on the footings. Drill holes in the bottoms of the stringers and set the stringers onto the bolts.

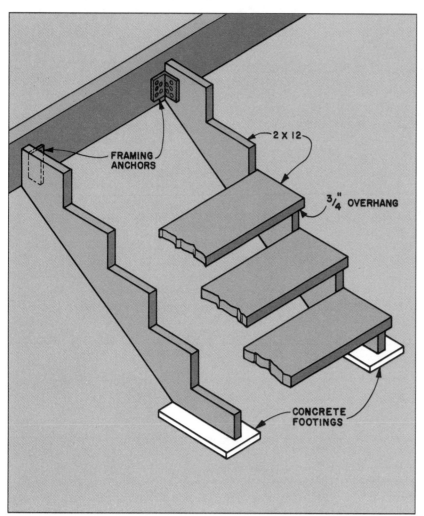

FRAMING ANCHORS

2 X 12

¾" OVERHANG

CONCRETE FOOTINGS

3 Use framing anchors to attach the stringers to the deck frame. Then cut the treads from 2 x 10 or 2 x 12 stock (depending on the run of each stair) and lay them across the stringers. Attach the treads to the stringers using 16d nails. Complete the installation of the stairs by adding the balusters and a railing.

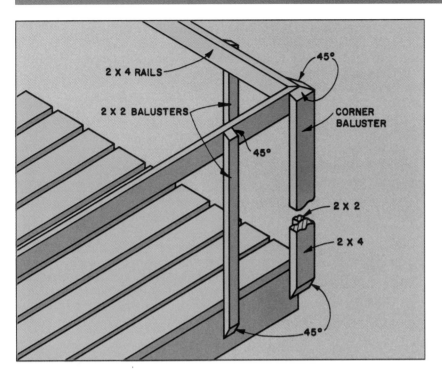

1 Cut the balusters from 2 x 2 stock and miter the top and bottom ends at 45°, so water will run off. To make the railing turn a corner, make an L-shaped baluster from a 2 x 4 and a 2 x 2. Double-miter the ends of the corner balusters, so they slope in two directions. Then, cut 2 x 4 stock for the rails, mitering the ends where the rails come together in the corners.

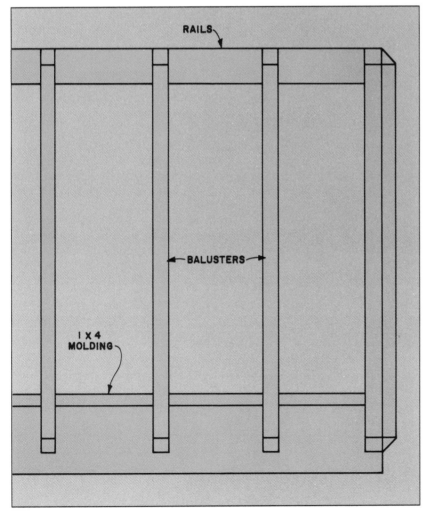

2 Lay out the balusters for one side of the deck, inside surface up. Lay the railing over the balusters, making sure the top edge of the rail is flush with the top ends of the balusters. Nail the rail to the balusters with 12d nails. With a helper, lift this railing assembly into place, and attach the bottom of the balusters to the outside joists or the facing board. Repeat the process for the other sides of the deck. Be sure to leave an opening for the stairs. When all the railing assemblies are in place, attach them to the house with framing anchors. To hide the end grain of the decking, nail 1 x 4 molding in between the balusters all along the edge of the deck.

APPENDIX 1 — ESTIMATING CONSTRUCTION MATERIALS

Lumber

Use a good set of scale plans to calculate the number of pieces you need. Since lumber is usually stocked in even lengths from 6 to 20 feet, be sure you have a way to get it to your work site. Most lumberyards will deliver. If you intend to haul it yourself, by car or 'small' vehicle, select short boards or have the lumberyard cut it to a manageable size. Allow for building with 'scraps' whenever possible. For example: Instead of ordering 14-foot boards for a 12½-foot span, use 8-foot boards and 4½-foot 'left over' boards — design permitting.

Sheeting Materials

This group of building materials includes plywood, particleboard, hardboard, waferboard, and fiberboard. These materials are used for subflooring, wall sheathing, and roof sheathing, so you should first learn the construction principles associated with each area. Using a set of plans, simply sketch out the number of 4 x 8 sheets needed. If you need to convert this figure to square feet, which is sometimes required by suppliers, simply multiply the number of sheets by 32.

Siding

For any rectangular wall area, multiply the height by the width (both figures rounded off to the nearest foot) to determine the total area. For a triangular wall area at a gable end, multiply the roof's rise by one-half of the span. A quick way to estimate the height of existing sided walls is to measure one shingle or board and multiply that measurement by the number of shingles or boards exposed on the wall.

With the total area calculated, next subtract the areas of all openings. To the resulting figure, add an additional 10 percent for error and waste. If your home has very steep gables or other details that require extensive cutting, add an additional 15 percent.

Convert this total figure to the number of sheets needed, if the siding is to be plywood. If your supplier needs to know the amount of siding needed expressed in linear or board feet, let him assist you in making the calculation.

Roofing

Roofing materials are usually sold in *squares,* an industry measurement equaling 100 square feet. When estimating for a simple gable roof, begin by calculating the surface area of each roof rectangle — multiplying length by width. Add all the areas together, then divide this figure by 100. Round the resulting total off to the next highest number. Use the same method as for siding to calculate more complicated 'non-rectangular' areas.

Insulation

Estimate insulation by square footage, using the same basic method as for siding. Multiply length by width to find the total area and then subtract for any openings. Once this figure is determined, multiply the resulting figure by 0.9 to allow for framing members that are spaced on 16-inch centers. If the framing members are on 24-inch centers, multiply the figure by 0.94.

Fasteners

Nails may be purchased by the box unless you need a lot of them; in this case, purchase by the pound. Estimating nails can be tricky but the easiest method is to count the number required in each unit to be nailed. Then multiply that figure by the number of units. Add 10 to 15 percent for error and waste.

To determine the number of framing connectors, bolts, or screws needed, simply count the number needed for each framing member and multiply by the number of members.

APPENDIX 2 — ESTIMATING CABINETMAKING MATERIALS

The most important aid in selecting and estimating cabinetmaking materials is the detailed project plan. With this in hand, you will be able to determine exactly what type material you need and also what

dimensions are needed. Remember to allow for the saw kerf plus cutting and sanding errors or problems. Also, make sure that you know exactly how the grain pattern will run.

APPENDIX 3 — BUDGETING CARPENTRY AND CABINETRY PROJECTS

After you have estimated the materials needed for your project, calculate how much each individual material will cost and then use the following basic budgeting sheet to figure the cost for the job.

Begin by comparing prices from several different suppliers, by phone, until you find the lowest price for equal quality. Try to settle on only one supplier and purchase everything at once, if possible. You might be able to take advantage of a contractor's discount if you're working with a contractor.

Most materials can be found at a building supply store, however large home improvement or hardware stores are good sources for hardware, both utility and decorative. Order wood and sheeting materials in standard sizes and in quantities 5 to 10 percent greater than your estimated needs. For sheeting materials, calculate the total square feet needed and multiply it by the price per square foot. For lumber, do the same but multiply it by the price per board foot.

Tools

Hand tools	$ _____
Portable power tools	$ _____
Stationary power tools	$ _____
Safety gear	$ _____
SUBTOTAL	$ _____

Fastening Materials

Nails	$ _____
Screws	$ _____
Bolts	$ _____
Framing connectors	$ _____
Adhesives	$ _____
SUBTOTAL	$ _____

Hardware

Hinges	$ _____
Door handles	$ _____
Drawer pulls	$ _____
Drawer-glide assemblies	$ _____
Shelf standards	$ _____
SUBTOTAL	$ _____

Construction Materials

Concrete	$ _____
Steel rods	$ _____
Cement block	$ _____
Vapor barrier	$ _____
Wire mesh	$ _____
Framing lumber	$ _____
Sheet material	$ _____
Building paper	$ _____
Siding	$ _____
Insulation	$ _____
Roofing material	$ _____
Underlayment	$ _____
Downspouts	$ _____
Gutters	$ _____
Flashing	$ _____
Drip edge	$ _____
Doors	$ _____
Windows	$ _____
Caulking	$ _____
SUBTOTAL	$ _____

Finish Materials

Wiring	$ _____
Plumbing	$ _____
Drywall	$ _____
Joint tape	$ _____
Joint compound	$ _____
Flooring	$ _____
Molding	$ _____
Abrasives	$ _____
Wood filler	$ _____
Finishes	$ _____
Cabinet and countertop materials	$ _____
Brushes and applicators	$ _____
SUBTOTAL	$ _____

GLOSSARY OF CABINETRY AND CARPENTRY TERMS

Abrasives. Materials such as sandpaper and steel wool used to smooth a workpiece for finishing.

Balloon framing. A framing method in which wall studs extend in one piece from foundation to roof, and floor joists are attached to wall studs.

Base shoe. Narrow molding used where the baseboard or a cabinet meets the finish floor.

Bearing wall. A wall used to support the weight of the structure.

Bird's mouth. A notch cut at the bottom of a rafter where it fits over the top plate.

Board foot. A traditional unit of measurement for pricing lumber; 1 square foot with 1 inch nominal thickness.

Brace. A piece of lumber applied to floor or wall; often used temporarily during framing.

Bridging. Pieces of wood or metal placed between joists or studs for reinforcement.

Building codes. Specifications and standards for the construction industry that protect the health and welfare of the public.

Building paper. A construction material that is used for protection against moisture and air filtration.

Butt joint. A joint that results when a flat end of one member is fitted to the face of another.

Casing. Molding around door and window openings.

Cleat. A strip of fastened wood used for the reinforcement or support of a shelf, cabinet, or other structure.

Collar beam. A connecting member used between rafters to reinforce the roof.

Coped joint. A process of cutting the end of one piece of molding at an interior angle to fit the molded face of another.

Counterbore. To bore a hole which measures the diameter of the screw head.

Countersink. To shape a screw hole so the head of a countersunk screw will finish flush with the surface.

Crawlspace. In houses with no basement, the shallow space between ground and first floor.

Cripple studs. The short studs at window and door openings.

Crosscut. To saw lumber or sheeting materials across the grain.

d. Penny. A traditional nail size.

Dado joint. A joint that results when the end of one member is fitted into the channel of another member.

Dimension lumber. Softwood lumber measuring at least 2 inches nominally thick; also called *light framing*.

Dowel. A small-diameter rod usually made of hardwood — often used for reinforcing joints.

Drawer glide assembly. Hardware fastened to the carcase and the drawer which supports the drawer and gives it its mobility.

Dryfitting. Preliminary assembling of wood members to check the fit without the use of permanent fasteners or glue.

Drywall. A material used for the interior finish of walls. Also called gypsum wallboard, plasterboard or sheet rock.

Eave. The roof overhang that projects beyond the exterior wall.

Elevation. An architectural drawing showing a head-on view of a wall or vertical surface.

Face frame. The framework attached to the front of a cabinet carcase.

Fascia. Trim used along the eave or gable end of a roof structure.

Fence. A stop, fixed or adjustable, against which a workpiece can be guided.

Fire block. A wood member placed within the structure of a wall to prevent the spread of fire or smoke through the air space.

Flashing. Material, usually metal, used at the roof and wall to prevent water penetration.

Footer. A masonry pad, usually concrete, for supporting a foundation wall or pier.

Foundation. The supporting portion of the structure, below the grade or below the first floor construction.

Framing members. The studs, joists, rafters, and other framing components that form the skeleton of a structure.

Frieze. A horizontal framing member that connects the siding with the soffit.

Furring strips. Pieces of wood secured to walls or other surfaces to serve as a fastening base for finish materials.

Gable. The triangular area of a structure that is formed by the pitch of the roof.

Girder. A horizontal beam for supporting interior joists or walls.

Grain. The configuration of the fibers in a piece of wood. Working with the grain is working parallel with the fibers. Working across the grain is working perpendicular with the fibers.

Green lumber. Unseasoned, undried lumber with a moisture content over 19 percent.

Groove. A channel cut with the grain in the edge or surface of the stock.

Gutter. A channel attached at the edge of a roof for carrying away rain water.

Hardboard. A sheeting material made of compressed wood fibers.

Hardwood. Wood from deciduous trees.

Header. A horizontal load-bearing support above window or door openings.

Jack rafter. A short rafter running from the ridgeboard to the hip or valley rafter, or running from the hip rafter to the top plate.

Jamb. The surface which lines the sides and top of window and door frames.

Joist. A horizontal framing member positioned on its edge for supporting floor or ceiling loads.

Jointing. Fitting wood members together with joints.

Kerf. The cut in a piece of wood made by a saw blade.

Kiln-dried lumber. Wood that has been dried in ovens, under controlled conditions, to remove excess moisture.

Ledger. A piece of lumber attached to a vertical member for the support of a joist or other horizontal member.

Lumber. Solid pieces of softwood or hardwood.

Miter joint. A joint that results when two wood members are cut at an angle other than 90°. The most common miter cut is 45°.

Molding. Specially milled wood strips used to protect edges, for decorative trim, and to hide areas where walls meet floors and cabinets meet walls.

Mortise. A cut-out in a wood member to fit a tenon or for fastening recessed hardware.

Mudsill. The lowest part of the structure that attaches to the foundation.

Nominal. In name only. Boards are sold in nominal sizes; thus they are smaller after being dried and planed smooth than their actual size.

Nosing. The rounded front edge of a stair tread.

On center. Indicates regular spacing between framing members; the distance from the center of one member to another. Also referred to as O.C.

Particleboard. A sheeting material made of compressed wood chips.

Pilot hole. A small hole drilled into wood for the purpose of avoiding splits when driving a screw or nail.

Pitch. In reference to the roof, the ratio of the total rise divided by the span.

Platform framing. A framing method in which each floor is built independently. The top plate of the floor below is the base for the joists of the floor above.

Plumb. Absolutely vertical.

Plywood. A material made of thin wooden layers (plies) all glued together with the grain of successive layers at right angles.

Purlin. A horizontal member supporting rafters or spanning between trusses for support of the roof covering.

Rabbet joint. A joint that results when the notched end of one member fits against the edge of another member.

Rafter. A framing member that extends from the top plate to the ridgeboard.

Rail. The horizontal member between two vertical members such as in a carcase frame or a frame-and-panel door.

Ribbon. A narrow member attached to studs or other vertical members for supporting joists or other horizontal members.

Ridgeboard. The horizontal member at the roof's peak to which the rafters are fastened.

Rip. To saw lumber or sheeting materials along their grain.

Rise. Stairs — The height of each riser. The total of all the individual risers in a set of stairs is the total rise of the staircase. Roofs — The vertical distance the ridge rises above the top plate at the center of the span.

Risers. The vertical boards between the stair treads.

Rough sill. The lowest framing member of a window or door opening.

Run. Stairs — The width of each stair tread. The total of all the individual runs in a set of stairs is the total run of the staircase. Roofs — One half the span.

Shakes. Hand split wood shingles.

Sheathing. Sheet materials such as plywood and particleboard that cover the studs and rafters.

Shim. A thin piece of material often wedged between gaps in members for establishing level or plumb.

Sill. The lowest part of the structure which rests on the foundation, also called the *mudsill*. Also, the lowest member of a door or window.

Slope. In reference to the roof, the rise in inches per foot of the roof's run.

Soffit. The underside of the roof overhang or the framing area above wall cabinets.

Softwood. Wood from coniferous trees.

Sole plate. The lowest horizontal member in wall construction — for supporting the studs.

Span. In reference to the roof, the distance between two opposing walls as measured from the outside of the top plates.

Stile. A vertical member in a face frame.

Stool. A flat piece of trim fitted over the window sill between the jambs.

Stops. Narrow wood strips for keeping doors and double-hung windows in place.

Story stick. A long strip of wood used to lay out and transfer measurements for determining the rise of a staircase, window and door openings, and the size of other framing members.

Stringers. The main support for a staircase; usually notched to accommodate the treads and risers.

Stud. The vertical framing members of a wall.

Subflooring. Plywood or boards placed on joists as a base for finish flooring.

Tail. In reference to roofing, the part of the rafter that extends beyond the structure.

Thread. The spiral section of a screw.

Toekick. The inset space along the bottom of a base cabinet.

Toenail. To drive a nail at an angle so as to join a vertical and a horizontal member.

Tongue and groove joint. A joint made by joining one member which has a projection to another member which has a groove or rectangular channel.

Treads. The horizontal surfaces of stairs.

Trimmer. In floor framing, a joist or beam into which a header is framed. In wall framing, a stud adjacent to studs at window and door openings for strengthening the sides of the opening and bearing the weight of the headers.

Truss. A framed unit, typically triangular, for supporting a load over a span.

Underlayment. Building material on which a finished floor is laid; also, material under finish roofing.

Vapor barrier. A building material, usually plastic sheeting, used for controlling moisture transmission; used in combination with insulation, it helps to prevent condensation.

Veneer. A thin wood sheeting either sandwiched with other sheets or, when it's decorative, laminated to the surface of inexpensive materials such as plywood.

A

Abrasives, 166
Awl, 11

B

Balloon framing, 66-67
Board foot, calculation, 28
Bolts, 31, 37
Bookcase, 156, 162-163
Brace, 12, 14
Building paper, 112-113

C

Cabinet
  base, 174-175
  carcase construction, 145-147
  countertop, 30, 175, 178-181
  door, 154-158
  drawer, 145, 149-153
  face frame, 147
  hardware, 155-156, 159
  molding, 175, 181
  planning, 144-145, 160-161
  shelf, 145, 148
  standard dimensions, 144
  top, 147
  wall, 175-177
  web framing, 151, 154-156
Carcase construction
  case, 146
  frame, 147
  panel, 146
Carpenter's level, 10-11
Carpenter's pencil, 10-11
Carpenter's square, 10
Chalk line, 11-12
Chisels, 12, 14
Clamping, 42-43, 50
Clamps
  bar, 15, 16
  C-clamp, 15, 16
  handscrew, 15, 16
Combination square, 11
Coped joint, 136-137
Counterbore, 54
Countersink, 54

D

Danger zone, 4-5
Deck
  building, 231-236
  designing, 231
  elevation drawing, 230
  railing, 237, 239
  stairs, 237-238
Dividers, 11-12
Door
  cabinet, 154-158
  casing, 116, 137-138
  components, 116, 186
  frame, 188
  framing, 83, 86

  prehung, 113, 115, 117
Dovetail jig, 17
Drawer
  butt joint, 149
  construction, 145, 149-151
  glide assemblies, 34, 151-153
  types, 150-151
  web framing, 151, 154-156
Drill bits, 16, 18
Drill, power, 16-17
Drywall
  installing, 124-127, 196, 202-203
  repairing, 126, 130
  taping, 128-129

F

Fiberboard, 111-112
Files, 15-16
Finishes
  cleanup, 170
  exterior, 116, 118-120
  interior, 159, 166-169, 171
Floor, 127
  opening, 79
  vinyl, 135
  wood, 132-134
Folding wooden rule, 10
Foundation
  building, 62-65
  elements, 60
  footer, 60-62
  foundation wall, 60-62
  posts and girders, 78
  slab, 60, 78
Framing, fasteners, 36-37, 95, 100

G

Glue, 30, 32
Gluing, 42-43, 50
Grain, 27, 41
Greenhouse window, 203-205
Gutters and downspouts, 119-121
Gypsum board, 111-112

H

Hammers, 12-13
Hardboard, 30
Hardware, 30-37, 156, 240
Hardwood, 28
Hinges, 32, 34-35, 155, 159

I

Insulation
  estimating, 240
  filler foam, 124
  installing, 125

J

Joinery
  butt joint, 44, 149
  dado, 44, 46-47

  dowels, 41, 49
  end lap, 44
  glue blocks, 41, 49
  groove, 44, 46-47
  lock joint, 48, 145, 150-151
  mid-lap, 48
  miter, 42, 44
  mortise and tenon, 48, 51
  rabbet, 44-45
  reinforcing, 41, 49
  splines, 41, 49
  tongue and groove, 48
Jointer, 19-20

K

Kerf, 12, 40

L

Level, 10
Lumber
  board foot calculation, 28
  defects, 27
  dimension lumber, 26
  estimating materials, 240
  grading, 26, 28
  grain, 27, 41
  hardwood, 28
  kiln-dried, 27
  knots, 27
  nominal, 28
  pitch pockets, 27
  pricing, 28
  separations, 27
  softwood, 26-28
  warpage, 27

M

Mallets, 12-13
Molding, 136-137, 175, 181

N

Nailing techniques, 51-53, 134
Nails, 31, 33

P

Particleboard, 29
Pilot hole, drilling, 52, 54-55
Planer, 19-20
Planes, 12, 14, 17-18
Planning
  architect, hiring, 70
  architectural symbols, 70, 75
  budgeting, 241
  building permits, 71
  building codes, 71
  cabinet, 144-145
  contractors, 72
  drawing plans, 70
  elevation drawing, 74
  estimating materials, 240
  plot plan, 73

skill levels, 72
Plaster, patching, 131
Plastic laminate, 30, 175, 178-180
Platform framing, 65, 67
Plumb bob, 10-11
Plumb, 10
Plywood
  characteristics, 28-29
  core materials, 29-30
  cutting, 41
  grading, 29
  sheathing, 104, 111-112
  siding, 116
  sizes, 29
  subflooring, 80, 82-83
Post and beam construction, 70

R
Rafter, stepping off, 68, 71
Railings, 226-227, 237-239
Rasps, 15-16
Repairing wood, 156, 164-166
Roller stand, 20-21
Roof
  components, 104
  drip edge, 104, 107
  estimating materials, 240
  fascia, 98
  flashing, 104, 106-107
  flat, 67
  frieze, 98
  gable, 67
  pitch, 68
  rise, 68
  run, 68
  shakes, 109-110
  sheathing, 100, 104-106
  shed, 67, 100
  shingles, 108-109
  slope, 68
  span, 68
  types, 67
  underlayment, 104, 106-107
Roof framing, 91
  bird's mouth, 68
  ceiling joists, 69, 91
  components, 68-69
  gable, 69, 93, 98-99, 101
  rafter, stepping off, 71
  ridgeboard, 69, 91
  soffit, 98, 100
  terminology, 68
  trusses, 70, 91-92, 94-95
Router, 17-18
  cutting dado, 46-47
  cutting groove, 46-47
  dovetail jig, 17
  guide, 46-47
  table, 17

S
Safety, 2-5, 7, 171
  checklist, 2-3

Sanders
  belt, 17-18, 20
  disc, 20
  orbital, 17-18
Sandpaper, 166
Saw blades, types, 41
Sawdust, control, 4-5
Sawhorses, 20-23, 177
Sawing techniques, 40-41
  avoid tear-out, 41
  bevel, 43
  crosscut, 40, 43
  miter, 42
  rabbet, 45
  rip, 40, 42
Saws
  backsaw, 12-13
  circular, 16, 18
  coping, 12-13
  crosscut, 12-13
  kerf, 12, 40
  keyhole, 12-13
  point, 12
  radial arm, 18-19, 43
  sabre, 16, 18
  set, 12
  reciprocating, 92
  table, 18-19, 42-43, 45
  table extension, 20-21
Screwdrivers, 14, 16
Screws, 31, 33, 36
Screws, covering, 55
Sheathing
  estimating materials, 240
  installing, 104-105
  types, 28-30, 100, 111-112
Shelf
  adjustable, 145, 148
  brackets, 34
  cleats, 148
  dowels, 148
  fixed, 145, 148
  pins, 148
  standards, 34, 145, 148
Siding, 109
  estimating materials, 240
  installing, 118
  types, 115-116
Skylights, 212-217
Softwood, 26-28
Stairs
  building, 221-223
  components, 220
  deck, 237-238
  designing, 220
  framing, 226
  installing, 224-226
  railing, 226-227, 237-239
  rise, 222
  run, 222
Steel tape, 10
Steel wool, 166
Story stick, using, 222
Stringer, laying out, 223
Subflooring

bridging, 61, 66-67, 80-81
building, 66-67
components, 61
joists, 61, 66, 79-80
girder, 80
installing, 80-83
materials, 80
mudsill, 60-61, 78
opening, 79

T
T-bevel, 10-11
Termite, prevention, 65, 67, 78
Try square, 11

V
Vapor barrier, 60, 65, 125

W
Wall framing, 82-91
  balloon, 66-67
  bearing wall, 65-66, 82-83
  corners, 89
  cripple studs, 65, 86-87
  door, 83, 86, 188
  fire blocks, 66
  header, 65, 86-87, 89
  platform, 65, 67
  rough sill, 65
  sole plate, 65
  straightening, 90-91
  top plate, 65, 91
  trimmer stud, 86-87
  raising, 88
  sheathing, 111-112
  stud, 65-66, 84, 86-87
  tying into, 91-93
  window, 83, 87, 189, 191-195,
    199-201
Window
  bay, 205-209
  casing, 113, 137, 139
  components, 113
  framing, 83, 87
  octagonal, 189, 193-196
  prehung, 113-115
  quarter-circular, 197, 202
  round, 189-192
  seats, 210-212
  semi-circular, 197, 202
  stationary, 197-201
Wood
  grain, 27
  products, 31
  repairing, 156, 164-166
Workshop
  planning, 4-7
  safety, 7, 171
  work triangle, 6

Y
Yankee drill, 14, 16